Statistical physics and thermodynamics

Statistical physics and thermodynamics

Professor B. Jancovici
Faculté des Sciences d'Orsay, France

Problems by
Yves Archambault

Translated from the French by
L. J. Carroll
University of Liverpool

A HALSTED PRESS BOOK

John Wiley and Sons
New York

Published in the U.S.A.
by **Halsted Press**
A Division of **John Wiley and Sons, Inc., New York**

ISBN 0 470 43965 3

Library of Congress Cataloging in Publications Data
Jancovici, Bernard.
Statistical physics and thermodynamics.
"A Halsted Press book."
Translation of *Physique Statistique et Thermodynamique*.
Bibliography: p.
1. Thermodynamics. 2. Statistical physics.
I. Title
QC311.J2813   536'.7   72-12614
ISBN 0-470-43965-3

Printed and bound in Great Britain.

Contents

Preface

The object of this book is to present a unified description of thermodynamics and statistical physics intended for undergraduates who are meeting the subjects for the first time.

The development of thermodynamics from foundations as simple as the two laws was a remarkable achievement on the part of the great pioneers Carnot, Kelvin, and Clausius. Even now the classical theory remains a model of rigour and elegance whose study constitutes a stimulating intellectual exercise. However the abstract character of such concepts as heat, internal energy, and entropy makes such an exercise difficult for beginners without a mechanical interpretation of these ideas. In particular the unfamiliar state function $\int \delta Q/T$ can appear as a somewhat mysterious entity.

Consequently rather than wait for *Part Two: Statistical Physics* to introduce the reader to the existence of atoms, I have chosen to employ such microscopic ideas immediately to clarify the development of the subject. However, after the introduction presented here, the study of an account of the classical approach would be most profitable. There are a number of excellent works on thermodynamics to which the reader is referred.

To present the microscopic point of view adopted here in a completely logical way, some apparently elementary concepts such as temperature would only be introduced at a rather late stage. In order to be able to discuss concrete physical phenomena immediately, I have not attempted to follow a purely deductive order; for example a provisional definition of temperature is given early in the book to await a further treatment of the subject later. An account following a more logical sequence is possible, but did not seem to me appropriate at the level of the present work.

The range of topics discussed is probably larger than can be handled in the first years. It is left to the reader to make a choice according to his interest and the time at his disposal.

It is a pleasure to acknowledge the help of all those colleagues who have supplied me with their comments on the preliminary draft of the book.

Bernard Jancovici

Preface to the English edition

Only one minor change has been made in the text. A number of errors in the numerical answers to the problems have however been corrected, thanks to the translator, L. J. Carroll, who has reworked all the solutions; I wish to acknowledge his care and patience.

B.J.

Introduction

Thermodynamics and statistical physics are concerned with properties of matter and more generally with systems whose behaviour involves the concepts of temperature and heat.

Matter possesses a microscopic structure in the form of molecules, atoms, electrons and nuclei. All the properties of matter including thermal properties must, in the final analysis, be capable of explanation in terms of the laws of classical or quantum mechanics, which these microscopic constituents obey. This is the point of view of statistical physics or statistical mechanics.

The viewpoint of thermodynamics, or more precisely classical thermodynamics as opposed to statistical physics, is to consider only the macroscopic properties of matter, i.e., its behaviour on our scale, and to establish relations among these properties without introducing the underlying microscopic mechanisms. For example, if the way in which a gas expands when its temperature is increased is known, thermodynamics enables the heat produced by the same gas when its pressure is increased to be calculated.

Throughout this book, we will adopt the microscopic point of view as systematically as possible rather than follow the historical order of development of the subject. Starting from microscopic considerations, the laws of thermodynamics will be established without too much concern for rigorous argument. From these laws, the purely macroscopic relations that make thermodynamics interesting on a practical level will be developed. Such relations, while only having a restricted application, are exact and can be stated without any knowledge of the detailed structure of the system being considered. The microscopic point of view however will provide a better understanding of the macroscopic laws as well as leading to results that cannot be obtained with classical thermodynamics.

Chapter 1 is a preliminary descriptive study of matter. Chapters 2 and 3 use statistical physics to establish the macroscopic laws of thermodynamics. Some developments and applications of these laws are discussed in chapter 4. Chapter 5 is devoted to questions of statistical physics going beyond macroscopic thermodynamics.

1. Description of the states of matter

1.1 Macroscopic variables. Equilibrium

A *system* is a body or collection of bodies confined in a region of space. For example, one gram of air enclosed in a balloon is a system. The air has a complex *microscopic* structure and the detailed description of its state brings into play a large number of variables, such as the position of every molecule, its speed, its direction, and so on. However, on the less detailed scale of our everyday world, the properties of the system are adequately described by a small number of *macroscopic* variables, such as volume, mass, pressure, temperature, index of refraction, and so on. Some of these variables, the volume or mass for example, are proportional to the amount of matter present; they are called *extensive variables*. Other variables, such as the pressure, temperature, and index of refraction, do not depend on the quantity of matter present; they are called *intensive variables*.

Starting from some arbitrary initial conditions, a system in general undergoes changes from a macroscopic point of view; there can be changes in the distribution of matter in the form of currents, chemical reactions, and so on. However, if the system is left to itself, it eventually reaches of its own accord a state of rest on a macroscopic scale, called a state of *equilibrium*. The macroscopic variables then have well-defined fixed values. This book will be concerned especially with the properties of matter in equilibrium, which are the simplest to study.

Among the macroscopic variables, we will be particularly interested in *volume*, *pressure* and *temperature*. Concerning pressure, we recall that a fluid exerts on an element of surface immersed in it a force normal to that element and proportional to its area; the force per unit area is called the pressure. In the SI system, the unit of pressure is the newton per square metre.[1] We know that the pressure of a fluid in equilibrium is governed

[1] Translator's note. In the original French edition, which used the MKSA system, the pascal was quoted as an alternative name for the newton per square metre. This name is not a part of the SI system.

by laws of hydrostatics and, strictly speaking, depends on the height of the point being considered. However, this effect is of little importance in a volume of fluid of the usual dimensions and we will neglect it. Thus the fluid will be characterized by a well-defined value of pressure. A solid can be subjected to a well-defined pressure by immersing it in a fluid, and this pressure will be that of the fluid surrounding it.

The concept of temperature needs a more detailed study which we will begin in the next section.

1.2 Introduction to the concepts of temperature and heat

We will be led to refine the concept of temperature throughout this book; the present section constitutes only an introduction.

Our first, subjective, impression of temperature comes from the sense of touch, which tells us how hot or cold a body is; the sense of touch enables us to say that one temperature is higher or lower than another. The change of temperature of a body is accompanied by variation of many macroscopic properties of the body, such as the volume, the electrical resistivity, and so on.

The sensation of temperature is a manifestation of a physical phenomenon on a microscopic scale: the irregular motion of the molecules which make up the body. This irregular motion becomes more energetic as the temperature is raised. With this in mind, we will attempt to define temperature as a parameter which is related, in a way which we will have to specify, to the magnitude of the irregular motion. A rather direct demonstration of this molecular activity is the existence of *Brownian motion*. If a drop of water, in which there is a suspension of fine, solid particles with diameters of the order of a micron, is observed under a microscope, the particles are seen to have an unceasing disordered motion. This is called Brownian motion after the botanist Brown, who discovered it in the last century but did not explain its origin. It occurs because the water molecules, which are not directly visible, are continually striking the solid particles from every direction, setting them in motion.

The sense of touch enables us to establish that, if two bodies with different temperatures are placed in contact, the hotter one cools and the colder one becomes warmer until the temperatures of the two bodies are equal. The fact that a change has taken place also manifests itself in the variation of a number of measurable quantities; for example, the hotter body contracts and the colder body expands. When the change is completed, the two bodies are said to be in *thermal equilibrium* with each other.

This equilibrium is a direct consequence of intermolecular collisions. The more energetic molecules of the hotter body collide with the less energetic ones of the colder body and transfer part of their energy to them. The energy transferred by these random molecular collisions is called *heat*. The transfer of heat ceases and thermal equilibrium is achieved, when the magnitude of the irregular molecular motion is the same for both bodies. What is to be understood by irregular motions of the same magnitude for molecules which may be of different types, will be specified later.

The rate at which the transfer of heat takes place depends on the nature of the materials considered. Some substances, such as cork or glass wool, only reach thermal equilibrium with another body very slowly. These substances enable us to make *adiabatic*[2] partitions which transmit practically no heat. Two bodies can be maintained at different temperatures for a long time if separated with such a partition. An adiabatic partition can also serve to isolate a body or collection of bodies thermally from the outside world. In contrast, with substances such as copper, the transfer of heat takes place quickly; a partition made of such a material is said to be *diathermic*.

Consider two bodies *A* and *B* which, placed in contact with body *C*, have shown by the absence of further change that they are both in thermal equilibrium with *C*. It is a matter of experience that *A* and *B* do not change if they are placed in mutual contact and are therefore in thermal equilibrium with one another. Thus:

Two bodies which are in thermal equilibrium with a third are in thermal equilibrium with each other.

This proposition constitutes the *zeroth law of thermodynamics* (the explicit statement of this law was made at a relatively late date and the 'first law' had already been assigned to the conservation of energy). This law is not self-evident. We will see in chapter 3 that it can be explained starting from the existence of temperature which has itself been defined with the help of microscopic considerations. Here however, we will consider the zeroth law as a fact of experience which allows the introduction of the concept of temperature.

In fact, we need only to be able to define temperature by means of a specific system that we will call a *thermometer*. The temperature of a body of any kind will then be defined as being the same as that of the thermometer, when the body and the thermometer are in thermal equilibrium. Two bodies at the same temperature in the sense that has just been indicated are then, as a consequence of the zeroth law, in thermal equilibrium with one another. Now we can see the significance of the law; without it the statement that the temperatures of two bodies were equal, would be the

---

[2] The adjective adiabatic is sometimes used in a different sense to the one which we have given it here.

product of measurements with a thermometer but would not necessarily indicate a special relation—thermal equilibrium—between these bodies.

It remains for us to define the temperature of the thermometer. To do this, a macroscopic property x, characteristic of the thermometer and dependent on the state of molecular motion, is employed and a fixed function of this property is taken for the temperature $t = f(x)$. For example, a mercury thermometer consists of a mercury-filled glass reservoir drawn out into a tube closed at one end. The thermometric property x is the length of the column of mercury in the tube and the temperature is a linear function

$$t = ax + b$$

of this length. The coefficients a and b can be determined from the temperature of two *fixed points*, agreed by convention. Ice in equilibrium with air-saturated water under a pressure of one atmosphere is a system with a well-determined temperature which is defined as zero; steam in equilibrium with water at a pressure of one atmosphere also has a well-determined temperature which is assigned the value 100. Since a linear relationship has been chosen between length and temperature, the temperature on the thermometer can be read directly by means of a scale consisting of a series of equal divisions. Thus a *centigrade* scale of temperature can be defined; temperature can be expressed in degrees centigrade on this scale.

The scale obtained in this way has an essentially arbitrary character. We are not bound to choose a linear relation between the thermometric property and the temperature. Even in limiting ourselves to a linear relation, we can make other conventions for the choice of coefficients. Finally and above all, there is nothing fundamental in the choice of the length of a column of mercury for the thermometric property; for example, the electrical resistance of a wire could just as well have been chosen. Compare the mercury thermometer with another thermometer filled with alcohol. The two instruments have been calibrated in such a way that both indicate 0 in melting ice and 100 in boiling water. The two thermometers will indicate the same reading at all other temperatures only if there is a linear relation between the lengths of columns of mercury and alcohol. There is no reason why this condition should be satisfied and in fact it is not; in general the two thermometers will therefore indicate different readings when in contact with the same body.[3]

As a standard scale of temperature, a particular scale has been chosen which can be defined starting from properties common to all gases when their densities are very low (they are then called perfect gases). We will

[3] In practice the difference is small, of the order of a tenth of a degree. However, more significant differences can be obtained with thermometers using other thermometric variables.

study this scale, called the *absolute scale*, in section 4 of the present chapter. It will appear in chapter 3 that the absolute scale has a special status and plays a fundamental role in thermodynamics.

1.3 Independent variables. The equation of state

The different macroscopic variables which characterize a system are not always independent. For example, we can arbitrarily fix the volume and the temperature of a fixed mass of fluid, and the pressure will then be determined. There exists therefore a functional relation between volume, pressure and temperature; any two of these quantities can be considered as independent variables, of which the third is a function. This relation is called the *equation of state*.

The macroscopic description of a solid is more complex; a solid can be governed by forces characterized by variables other than a simple pressure and it follows from this that deformations in a solid are characterized by variables other than a simple volume. We will content ourselves here with studying some simplified situations. We will at least be able to speak of an equation of state between the volume, pressure and temperature for a solid that is governed only by a hydrostatic pressure. We will also be able to consider the case of a wire; this will be described by its length, its tension and its temperature, which, as long as the tension is not too large, are related by an equation of state.

Hysteresis effects are another source of complication. There is *hysteresis* when the properties of the system depend on its past history. For example, if a metallic wire is repeatedly pulled too vigorously, the lengthening produced by the same tensile force increases with each application; the length of the wire is no longer a well-defined function of its tension and temperature and there is no longer an equation of state. We will exclude from our study systems in which hysteresis effects occur. In any case, these systems should not, strictly speaking, be considered as systems which have reached equilibrium; they change slowly in the course of time although often in an imperceptible way.

1.4 Perfect gases

A gas has particularly simple physical properties, when it is extremely rarefied. The ideal state, towards which all gases tend when their densities are very low, is called the *perfect gas*.

KINETIC THEORY OF PERFECT GASES

We are going to study perfect gases and establish their equation of state starting from microscopic considerations.

The model. A *perfect gas* is, by hypothesis, a gas of very low density; this means that the mean distance between the molecules is large compared with the range of the intermolecular forces (which is of the order of 1 Å, i.e., 10^{-10} m). The molecules therefore only interact occasionally with one another; a perfect gas can be represented as a collection of point molecules with no appreciable interaction between them moving freely inside the enclosure which contains the gas. Under usual conditions, the molecules have speeds of the order of hundreds of metres per second.

Pressure. The pressure which a gas exerts on the walls of its enclosure is due to the collisions of the molecules with these walls. The pressure is naturally greater, when the molecules have more energy. We will establish an exact relation between the pressure p and the kinetic energy of translation, U_t of the molecules.

Consider (Fig. 1.1) an element of the wall, of surface area S, and take for the x axis the normal to that element pointing outwards. We assume that the wall is perfectly reflecting.[4] A molecule which strikes the wall with an

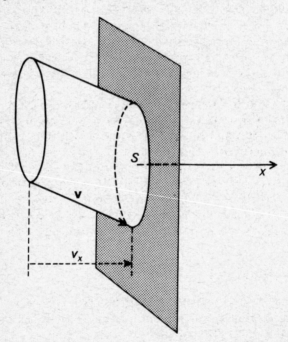

Figure 1.1. A collision between a molecule and a wall of its enclosure.

---

[4] This hypothesis of a perfectly reflecting wall is made here for the sake of simplicity. At the price of slightly more complicated reasoning, it is possible to take into account the fact that the wall has, on a microscopic scale, a complex structure which is different from that of a plane.

x component of velocity, v_x, (v_x is necessarily positive so that there may in fact be a collision) is reflected with an x component of velocity, $-v_x$; the other components of velocity are unchanged. Thus in the collision, the molecule, assumed to have mass m, increases its momentum perpendicular to the wall by an amount equal to $-2mv_x$. Assume for the moment that every molecule has the same velocity $\mathbf{v}$. If there are v collisions per unit time on the element of the wall, this element imparts momentum $-v2mv_x$ to the gas. According to the fundamental equation of dynamics, $\mathbf{F} = d\mathbf{P}/dt$, the force $\mathbf{F}$ exerted by the element of the wall on the gas is equal to the increase in the momentum of the gas per unit time, $d\mathbf{P}/dt$. This increase is directed along the x axis and has the value $-v2mv_x$. Conversely, the force exerted by the gas on the element of the wall is $v2mv_x$; it is perpendicular to the wall and directed outwards.

It is possible to calculate v by noticing that the molecules that strike the element of the wall in unit time are contained in an oblique cylinder, of base S and generator $\mathbf{v}$, whose volume is Sv_x. If there are n molecules per unit volume—n is called the *numerical density*—we have

$$v = nSv_x$$

The force exerted on the element of the wall is

$$v2mv_x = 2nmv_x^2S$$

and the pressure, the force per unit area, is

$$p = 2nmv_x^2$$

In the real world, every molecule does not have the same velocity $\mathbf{v}$. The pressure p is due to the average effect of molecules with a range of velocities. To account for this, it is necessary to replace v_x^2 by the *mean value*[5] $\overline{v_x^2}$ in the preceding formula and, since only half the molecules—those which have a positive v_x—contribute to the collisions, to replace the total numerical density n by $n/2$. Thus we find

$$p = nm\overline{v_x^2}$$

As the gas is isotropic

$$\overline{v_x^2} = \overline{v_y^2} = \overline{v_z^2} = \tfrac{1}{3}(\overline{v_x^2} + \overline{v_y^2} + \overline{v_z^2}) = \tfrac{1}{3}\overline{v^2}$$

Furthermore by definition

$$n = \frac{N}{V}$$

[5] Let f_i be a quantity associated with the ith molecule and let N be the total number of molecules. The mean value of f is by definition $\bar{f} = (1/N)\sum_{i=1}^{N} f_i$.

where N is the total number of molecules and V is the volume of the gas. Then

$$p = (N/V)\tfrac{2}{3}\tfrac{1}{2}m\overline{v^2}$$

We know the total kinetic energy of translation to be

$$U_t = N\tfrac{1}{2}m\overline{v^2}$$

and hence obtain the desired relation

$$pV = \tfrac{2}{3}U_t$$

known as *Bernoulli's formula*.

Temperature. The properties of perfect gases enable a particularly simple and convenient scale of temperature, called the *absolute temperature scale*, to be defined.

Since the temperature is a measure of the magnitude of the irregular molecular motion, we can define it as a quantity T proportional to the mean kinetic energy of translation of a molecule

$$\tfrac{1}{2}m\overline{v^2} = \frac{U_t}{N} = \tfrac{3}{2}kT$$

k is a constant, known as Boltzmann's constant, which will be determined by the units convention; the factor 3/2 is introduced for convenience in order to simplify the subsequent formulae.

By using Bernoulli's formula, *the equation of state for a perfect gas* can be obtained immediately

$$pV = NkT.$$

This relation can be considered as a macroscopic definition of absolute temperature, equivalent to the preceding microscopic definition. For a gas of fixed volume V, the pressure p is a thermometric property and the temperature T is defined as directly proportional to this pressure.

This definition of temperature is only acceptable because, if any two perfect gases are in thermal equilibrium with each other, the mean kinetic energy of translation of a molecule is the same for both gases—there is said to be *equipartition of energy*—and the parameter T, determined from this energy, is the same for both gases.[6]

It is not until section 3.4 of this book that we will be in a position to give a rigorous explanation of the equipartition of energy. We will content our-

[6] It is to obtain a T having this property that T has been defined as related by the universal constant k to the kinetic energy of translation *per molecule* rather than to the energy per unit mass or unit volume for example.

selves here with no more than making equipartition plausible, by appealing to the following fundamental concept which will be expressed in a more precise way later: the collisions[7] between molecules tend to distribute the energy among the molecules in a random manner and a system reaches equilibrium when, on a microscopic scale, it is in a state as disordered as possible.

Suppose to begin with that the two gases are mixed in the same container and consider a collision between a molecule of the first gas, mass m_1 and velocity $\mathbf{v}_1$, and a molecule of the second gas, mass m_2 and velocity $\mathbf{v}_2$. During the collision, assumed elastic, the total momentum $m_1\mathbf{v}_1 + m_2\mathbf{v}_2$ and the total energy $\frac{1}{2}m_1v_1^2 + \frac{1}{2}m_2v_2^2$ are conserved. This is equivalent to saying that the total momentum and the modulus $|\mathbf{v}_1 - \mathbf{v}_2|$ of the relative velocity are conserved.[8] On the other hand the direction of the relative velocity changes during the collision. The idea that collisions generate disorder, leads us to conclude that, when equilibrium is established, the relative velocity and the total momentum of a pair of molecules are oriented at random relative to one another on the average, i.e., on the average their scalar product is zero

$$\overline{(\mathbf{v}_1 - \mathbf{v}_2)\cdot(m_1\mathbf{v}_1 + m_2\mathbf{v}_2)} = 0.$$

Following the same line of thought, we also conclude that in the disordered molecular state that exists in equilibrium the velocities $\mathbf{v}_1$ and $\mathbf{v}_2$ are oriented at random relative to one another; on the average, their scalar product is zero

$$\overline{\mathbf{v}_1 \cdot \mathbf{v}_2} = 0$$

Combining the two preceding equalities yields the result

$$\tfrac{1}{2}m_1\overline{v_1^2} = \tfrac{1}{2}m_2\overline{v_2^2}$$

Thus there is equipartition of kinetic energy of translation between the molecules of the two mixed gases.

---

[7] If the molecules were strictly points and did not interact, there would be no collisions between them. In fact, the perfect gas is only an ideal situation and there is always an interaction between the molecules. We assume that it is small enough not to contribute appreciably to the total energy, but sufficient to allow the exchange of energy between the molecules.

[8] The identity

$$\frac{m_1 m_2}{2(m_1 + m_2)}(\mathbf{v}_1 - \mathbf{v}_2)^2 = \tfrac{1}{2}m_1 v_1^2 + \tfrac{1}{2}m_2 v_2^2 - \frac{(m_1\mathbf{v}_1 + m_2\mathbf{v}_2)^2}{2(m_1 + m_2)}$$

can be used. The left-hand side, which is moreover the energy in the centre of mass system, is conserved during the collision since the right-hand side is constructed from quantities which are conserved.

9

When the two gases are in different containers separated by a diathermic partition, there are no longer direct collisions between the molecules of one gas and those of the other. Thermal equilibrium is established by means of the molecules of the wall and the situation is more complicated to analyse. As we will return to the topic in section 3.4 from a much more general point of view, we will accept the result here without proof: there is still equipartition of energy between the molecules of one gas and those of the other.

Thus we can define absolute temperature starting from any perfect gas.

UNITS AND CONSTANTS

Some systems have a well-defined and reproducible temperature, independent in particular of the masses of the constituents; these systems are said to define *fixed points* of temperature. For example, a system consisting of ice, water, and water vapour yields a fixed point, *the triple point* of water. The perfect gas law

$$pV = NkT$$

defines the absolute temperature T once the constant of proportionality, or what amounts to the same, the temperature of a particular fixed point, has been chosen. *By convention* the temperature 273·16 is assigned to the triple point of water. This choice fixes the unit of absolute temperature, which is known as the degree Kelvin (K). The absolute temperature can therefore be determined from the pressure p of a perfect gas, whose volume is kept constant, by the relation

$$T = 273 \cdot 16 \frac{p}{p_t}$$

where p_t is the gas pressure at the triple point of water.

For commonplace applications, *the Celsius temperature t* (often called the centigrade temperature), *determined* from the absolute temperature by the relation

$$t = T - 273 \cdot 15$$

is used. The degree Celsius (°C) is therefore equal to the degree Kelvin but the zero of the scale is displaced. It can be seen that the triple point of water is 0·01°C.

The temperature of melting ice under atmospheric pressure is a fixed point; this temperature is very nearly equal to 0·00°C. The temperature of steam in equilibrium with water at atmospheric pressure is another. This temperature is almost equal to 100·00°C. In fact before 1954 these temperatures were, by definition, exactly 0 and 100 on the centigrade scale and the displacement of the origin necessary to obtain the absolute scale had to be determined by measurement, which led to an embarrassing inaccuracy at low temperatures. The present definitions, which consist of assigning by convention the temperatures $273 \cdot 16\,K$ and $0 \cdot 01°C$ to the triple point of water, were chosen in such a way that the scales thus defined agreed as closely as possible with the old scales.[9]

The choice of the degree Kelvin as the unit of absolute temperature determines Boltzmann's constant k. The macroscopic laws are usually expressed in terms of a quantity of matter equal to one *mole* (or *gram molecule*). A mole is, by definition, the quantity of matter which contains N_0 molecules, N_0 being Avogadro's number. Since 1960 the *unified scale*, based on an atomic mass of 12 g for the carbon 12 isotope, has been adopted for the atomic masses. Avogadro's number is then the number of molecules contained in 12 g of C^{12}; its value is

$$N_0 = (6 \cdot 02252 \pm 0 \cdot 00009) \times 10^{23}$$

For a mole of gas, the perfect gas law is

$$pV = RT$$

where $R = N_0 k$ is the *gas constant*. By measuring the quantity pV/T, it is found that

$$R = 8 \cdot 31434 \pm 0 \cdot 00035 \, JK^{-1}$$

The value of Boltzmann's constant is then

$$k = R/N_0 = (1 \cdot 38054 \pm 0 \cdot 00006) \times 10^{-23} \, JK^{-1}$$

MACROSCOPIC PERFECT GAS LAWS

The general equation of state of a perfect gas

$$pV = NkT$$

incorporates several different laws which are customarily given their own special names.

At constant temperature, the product pV is a constant; this is known as *Mariotte's* or *Boyle's* law.[10]

---

[9] In the new convention the interval 0–100°C does not play a fundamental role; because of this, the term centigrade scale must be abandoned and Celsius scale used.

[10] Translator's note: Boyle's law is the more common title in English text-books.

By definition pV is proportional to T. If the pressure is kept constant and the volume at 0°C (273·15 K) is denoted by V_0 then

$$V = \frac{V_0 T}{273 \cdot 15} = V_0 \left(1 + \frac{t}{273 \cdot 15}\right)$$

Similarly, if the volume is kept constant and the pressure at 0°C denoted by p_0 then

$$p = p_0 \left(1 + \frac{t}{273 \cdot 15}\right)$$

These are the *laws of Gay-Lussac*, which in fact express no more than the definition of temperature starting from the perfect gas.

With the pressure, temperature and number of molecules fixed, the volume is independent of the nature of the gas. In particular at a temperature of 0°C ($T = 273 \cdot 15$ K) and normal atmospheric pressure ($p = 101\ 325\ \text{Nm}^{-2}$), it is found that the volume RT/p of one mole of any perfect gas has the value

$$V = (22 \cdot 4135 \pm 0 \cdot 0009) \times 10^{-3}\ \text{m}^3$$

This is *Avogadro's law*.

Finally we consider a mixture of several perfect gases. Since their molecules strike the walls independently, their contributions to the total pressure add together. If the number of molecules of each type are N_1, N_2, etc., then

$$pV = N_1 kT + N_2 kT + \ldots$$

This is known as *Dalton's law*. The pressure of a mixture of gases is equal to the sum of the partial pressures which the gases would exert separately, if each occupied the total volume on its own.

1.5 Thermometry

Thermometry is the measurement of temperature.

THE GAS THERMOMETER

Absolute temperature is defined by means of the perfect gas law. In measuring laboratories, gas thermometers are used to determine a number of fixed points. The absolute temperature is proportional to the volume of a perfect gas if the pressure is held constant, or to the pressure if the volume is held constant. In practice, a constant-volume thermometer is employed. In principle a perfect gas should be used, i.e., the pressure should be very low, but it would be difficult to make precise measurements under these conditions. For this reason a thermometer filled with gas at a pressure close to atmospheric is preferred; the filling pressure can be varied and the measurements extrapolated to low pressures.

The normal thermometer consists essentially of a platinum reservoir with a volume of about a litre, filled with hydrogen and connected to a mercury manometer (Fig. 1.2). The volume of hydrogen can be held constant. The

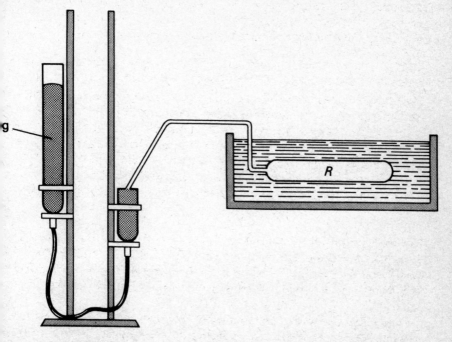

Figure 1.2. Simplified diagram of a gas thermometer. The reservoir R that contains the gas is immersed in a bath whose temperature is to be measured. The volume of the gas is maintained constant by adjusting the left-hand tube of the manometer to keep the mercury in the right-hand tube at the same level. The manometer enables the pressure of the gas to be determined (from A. L. King, *Thermophysics*, W. H. Freeman and Company, 1962).

reservoir is placed successively in thermal contact with the two systems A and B whose temperatures are to be compared. The corresponding hydrogen pressures p_A and p_B are measured and the ratio p_B/p_A calculated. The measurements are repeated several times with different quantities of hydrogen in the reservoir and the limit of the ratio as the quantity of hydrogen tends to zero is determined by extrapolation. The ratio of the absolute temperature of the systems A and B is

$$\frac{T_B}{T_A} = \lim \frac{p_B}{p_A}$$

The temperatures of various fixed points can be found by comparing them with the temperature of the triple point of water, known by definition to be 273·16 K.

For temperatures below 80 K, the hydrogen is too near its liquefaction point and is replaced by helium; temperatures as low as 1 K can then be reached. At temperatures above 600 K, the hydrogen would diffuse through the walls of the reservoir and for this reason it is replaced by nitrogen which can be used up to 1 400 K.

PRACTICAL THERMOMETERS. THE INTERNATIONAL SCALE

The gas thermometer is of little practical use and in fact serves only to determine certain fixed points. In practice, more manageable instruments, calibrated by reference to these fixed points, are employed. From such instruments, calibrated according to well-defined standards, *the international scale*, which is practically identical to the Celsius scale, is obtained.

For example, between 0 and 630°C, the resistance R of a platinum wire is well represented as a function of the Celsius temperature t by the formula

$$R = R_0(1 + at + bt^2)$$

The coefficients R_0, a, and b can be adjusted so that the formula is correct at the fixed points 0°C (the melting point of ice), 100°C (the boiling point of water), and 444·60°C (the boiling point of sulphur). The above formula then defines the international temperature throughout the interval from 0 to 630°C.

From -190 to 0°C the international temperature is defined by the same platinum resistance thermometer, governed by the formula

$$R = R_0[1 + at + bt^2 + c(t - 100)t^3]$$

where the additional coefficient c is found from the fixed point $-182·97$°C (the boiling point of oxygen).

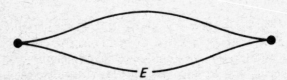

Figure 1.3. A thermocouple consists of two dissimilar metal wires welded together at their ends to form a closed circuit. If the junctions are at different temperatures, an electromotive force E that can be measured is developed in the circuit.

From 630 to 1 063°C a thermocouple (Fig. 1.3) consisting of a wire of platinum and a wire of platinum–rhodium alloy is utilized. With one junction fixed in melting ice, the electromotive force E of the thermocouple is related to the temperature of the other junction by the formula

$$E = a + bt + ct^2$$

The coefficients a, b, and c are determined from the fixed points 630·50°C (the melting point of antimony), 960·80°C (the melting point of silver), and 1 063·00°C (the melting point of gold).

Liquid thermometers—mercury or alcohol for example—are commonly used in everyday applications. As we have pointed out previously, such an instrument graduated in equal divisions defines its own characteristic scale of temperature, which does not coincide exactly with the perfect gas scale. By using a table of appropriate corrections, which are of the order of a tenth of a degree, the absolute temperature may be determined within known limits of precision. Nevertheless, a liquid thermometer is unsuitable for precise measurements, particularly since the volume of the glass envelope changes with age.

1.6 Real fluids

Isotherms

Clapeyron's representation. The perfect gas law only applies to real fluids asymptotically in the region of very low densities.

Outside this region, the equation of state, which depends on the pressure, volume, and temperature, is more complicated and can no longer be expressed precisely by a formula.

A useful graphical representation, *Clapeyron's representation*, consists of plotting curves which show the pressure p as a function of the volume V for a given temperature; these curves are called *isotherms*. Isotherms corresponding to several temperatures can be plotted on the same graph. A system of curves is obtained as illustrated in Fig. 1.4.

Starting from the perfect gas state, i.e., from large values of V, we will follow the behaviour of the fluid as it is compressed at constant temperature. There are two cases to consider depending on this temperature:

(*a*) If the temperature is above a certain value, called the critical temperature T_c, the pressure of the gas increases *indefinitely* without any singularities; p is a continuous, well-behaved function of V. However, the gas does not obey Boyle's law $p = \text{constant}/V$, except at very large values of V.

(*b*) If the temperature is below the critical value T_c, the phenomenon of *condensation* is observed. Below a certain volume V_g, the gas begins to liquefy; as V continues to decrease, more and more of the gas changes to the denser liquid state until at volume V_l there is nothing but liquid.

15

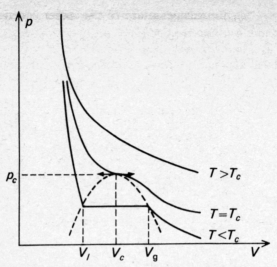

Figure 1.4. A family of isotherms in Clapeyron's representation. The heavy broken line is the saturation curve.

Throughout the region in which the gas is liquefying (V_g to V_l), the pressure remains constant; this horizontal part of the isotherm may be called the isobaric region of liquefaction. In this region, the gas and liquid densities remain constant, but the relative proportions of gas and liquid vary. Because of their different densities, the gas and the liquid are separated by gravity; the liquid occupies the bottom of the container with the gas above it,[11] the surface between them being a horizontal plane. Finally, if the liquid is compressed further to volumes less than V_l, the pressure begins to rise rapidly with decreasing V.

The critical temperature represents the dividing line between these two cases; the isotherm T_c does not show an isobaric region but a point of inflection with a horizontal tangent, which is called the *critical point*. The co-ordinates (V_c, p_c) of this point are the critical volume and pressure.

As the temperature is varied, the end points of the isobaric region of liquefaction describe the saturation curve, represented by the heavy broken line on the diagram. This curve specifies the region of co-existence, points inside the curve representing states in which gas and liquid co-exist.

The critical temperature depends on the gas under consideration. For example, it is 152°C for butane. Thus it is possible to liquefy butane at normal temperatures by compression; butane stored in cylinders is in the liquid state.

[11] In conditions near liquefaction the gas is often called a vapour.

On the other hand the critical temperature of nitrogen is very low ($-147°C$) and thus cylinders of compressed nitrogen contain gas. If nitrogen is to be kept as a liquid, it must be maintained at a low temperature in a suitably insulated container (a Dewar vessel).

Amagat's representation. It is often convenient to draw the system of isotherms using Amagat's representation; an isotherm then represents the product pV as a function of the pressure p for a given temperature. This representation makes the deviations from the perfect gas law particularly conspicuous. For a perfect gas at a fixed temperature, pV is constant and Amagat's isotherms are simply lines parallel to the p axis. For a real fluid, the set of curves indicated in Fig. 1.5 is obtained.

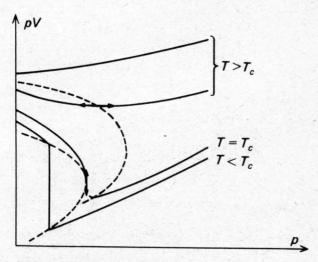

Figure 1.5. A family of isotherms in Amagat's representation. Boyle's curve and the saturation curve are shown as heavy broken lines.

When the pressure tends to zero, the gas tends to the perfect gas state. According to the perfect gas law, the ordinate pV is proportional to the absolute temperature T.

At very high temperatures, pV is a monotonic increasing function of p; the fluid is less compressible than a perfect gas. At lower temperatures, pV as a function of p has a minimum. In the neighbourhood of this minimum, pV varies little and the fluid behaves in the same way as a perfect gas, though this stationary value of pV is much lower than NkT. The locus of the minima of pV is a parabola-like curve, called *Boyle's curve*.

17

At the critical temperature, the isotherm displays a point of inflection with a vertical tangent. At temperatures below the critical temperature, each isotherm has a vertical region corresponding to liquefaction. The end points of this region form the saturation curve.

CONTINUITY OF THE LIQUID AND GASEOUS STATES

The shape of the system of isotherms is such that a continuous transition between the gaseous and liquid states is possible.

Let us consider for example the states A and B of Fig. 1.6. The transition from A to B may be made along isotherm $ACDB$ by simply compressing

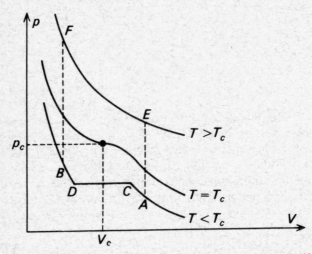

Figure 1.6. The continuous transformation between the gaseous and the liquid states.

the fluid at constant temperature. This transition goes through region CD, where the discontinuous process of liquefaction occurs. In region CD where the liquid and gas are quite distinct, the states have two phases. A is a gaseous state while B is a liquid state.

However, it is also possible to go from A to B following a path such as $AEFB$ which passes round the critical point. The transition from A to E is achieved by heating at constant volume, that from E to F by compressing at constant temperature and that from F to B by cooling at constant volume. Point B is reached without any property of the fluid passing through a discontinuity, the fluid remaining homogeneous throughout the transition.

This possibility of continuous transition from state A to state B makes the terms liquid and gas somewhat imprecise. There is no natural criterion which enables the distinction between liquid and gas to be made. If path *AEFB* is taken, where does the gas end and the liquid begin? Because of this ambiguity, it is convenient to speak merely of the *fluid state* without further qualification.

INTERMOLECULAR FORCES

To explain the properties of real fluids, it is essential to take account of the intermolecular forces.

The force between two molecules is strongly repulsive at short range and attractive at long range.[12]

The repulsion at short range is a consequence of the Pauli principle: two electrons cannot occupy the same quantum state. This restriction prevents the mutual penetration of the electron clouds of the two molecules.

The attraction at long range results from the interaction between the electric dipole moments carried by the molecules. While *polar* molecules, HCl for example, have a permanent electric dipole moment, we will be especially interested in the more simple case of *non-polar* molecules, i.e., those which have no permanent electric dipole moment. Molecules of hydrogen or of the monatomic noble gases are examples of these. Non-polar molecules however can be polarized; they can acquire an induced electric dipole moment when they are acted on by an electric field. Moreover, it is only the mean value of the moment—in the quantum mechanical sense— that is zero for a non-polar molecule; the instantaneous configurations of the nuclei and the electrons generate a non-zero moment fluctuating rapidly. When two non-polar molecules are brought near one another, each is polarized by the action of the electric field created by the instantaneous dipole moment of the other. The interaction between the instantaneous moment of one molecule and the induced moment of the other gives rise to an attractive force called a *Van der Waals force*. It can be shown that this force arises from a potential of the form $-1/r^6$, where r represents the distance between molecules.

The interaction between two non-polar molecules is well represented, particularly in the case of the noble gases, by a *Lennard-Jones potential*. The mutual potential energy of a pair of molecules as a function of their separation r is given by

$$\phi(r) = 4\varepsilon \left[\left(\frac{\sigma}{r} \right)^{12} - \left(\frac{\sigma}{r} \right)^6 \right]$$

ε and σ are parameters. By way of an example, we quote the values for argon

[12] Of course, as the total electric charge of a molecule is zero, the law of force between two molecules is not Coulomb's law!

(ε is an energy, and it is useful to quote ε/k in degrees):

$$\varepsilon/k = 119 \cdot 8 \text{ K}; \quad \sigma = 3 \cdot 405 \times 10^{-10} \text{ m}$$

The function $\phi(r)$ is shown in Fig. 1.7. In $\phi(r)$ the repulsive term $1/r^{12}$ represents the effect of the Pauli principle. The choice of the exponent 12 has been made empirically and has no particular theoretical justification. Alternatively, the repulsion is sometimes treated by considering the molecules more simply as impenetrable or 'hard' spheres. The attractive term $-1/r^6$ represents the Van der Waals force.

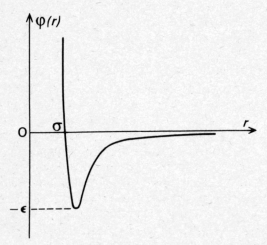

Figure 1.7. The interaction potential between two molecules.

The forces between polar molecules are more complicated; they depend on the orientations of the molecules. However, the general picture remains the same; there is repulsion at short range and attraction at long range.

Van der Waals' equation

A number of equations of state, empirical and theoretical, have been proposed to describe real fluids. Up to the present time, there is none which is in precise quantitative agreement with experiment at all temperatures and densities. By way of compensation it has been known for many years that Van der Waals' equation, which is based on molecular considerations, accounts correctly for the *qualitative behaviour* of the isotherms. It is not the ideal equation, but it has the important merit of being particularly simple. We will establish this equation.

For a perfect gas, we had the law

$$P = \frac{NkT}{V}$$

The pressure tends to infinity as the volume tends to zero.

For a real fluid, the intermolecular repulsion will oppose the compression of the fluid to an arbitrarily small volume. If the molecules are considered as hard spheres, the pressure will become infinite when the molecules are packed tightly together. They then occupy a volume Nb, where b is of the order of magnitude of the volume of one molecule. This effect can be approximately accounted for by replacing the perfect gas law by

$$p = \frac{NkT}{V - Nb} = \frac{kT}{v - b}$$

where $v = V/N$ is the volume per molecule.

Now we must account for the attractive part of the interaction. When a molecule approaches the wall of the container, it is decelerated by the attraction of the other molecules and the pressure on the wall is reduced. If the presence of that molecule has a negligible effect on the distribution of the others, the attraction felt by that molecule is directly proportional to n, the number of molecules per unit volume.

Since the number of molecules striking the wall is also proportional to n, the decrease in pressure is of the form $an^2 = a/v^2$, where a is a constant.

The combination of the effects of the repulsive and attractive forces leads to the equation of state obtained by Van der Waals

$$p = \frac{kT}{v - b} - \frac{a}{v^2}$$

For one mole, taking the number of molecules equal to Avogadro's number, we have

$$p = \frac{RT}{V - B} - \frac{A}{V^2}$$

where we have set $B = N_0 b$, $A = N_0^2 a$.

Let us examine (Fig. 1.8) the shape of the isotherms produced by Van der Waals' equation. At high temperatures, the first term $RT/(V - B)$ dominates the pressure p; the pressure is a decreasing function of V. On the other hand at low temperatures, the second term is equally important, and, as a function of V, p is found to first decrease, then increase and finally decrease once more. The transition from one type of behaviour to the other takes place at some temperature T_c; at this temperature the isotherm has simply a point

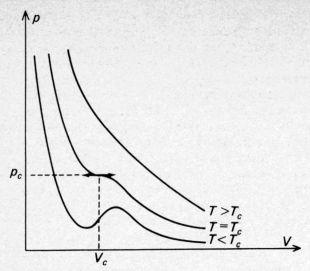

Figure 1.8. Isotherms calculated from Van der Waals' equation.

of inflection with a horizontal tangent. It is simple to express this critical temperature T_c, as well as the coordinates of the point of inflection (the critical volume V_c and the critical pressure p_c), as a function of the parameters of Van der Waals' equation. At the critical point we have

$$\frac{\partial p}{\partial V} = -\frac{RT_c}{(V_c - B)^2} + \frac{2A}{V_c^3} = 0$$

and

$$\frac{\partial^2 p}{\partial V^2} = \frac{2RT_c}{(V_c - B)^2} - \frac{6A}{V_c^4} = 0$$

We use partial derivatives because p is also a function of T. From these relations we conclude

$$T_c = \frac{8A}{27BR}; \qquad V_c = 3B$$

The corresponding value of p is

$$p_c = \frac{A}{27B^2}$$

It will be seen that

$$\frac{RT_c}{p_c V_c} = \frac{8}{3}$$

22

The experimental values should satisfy this relation if Van der Waals' equation is exact. In fact the ratio RT_c/p_cV_c often has a somewhat different value (1·10 for mercury vapour; 4·35 for water). These discrepancies give an indication of the limited validity of Van der Waals' equation from the quantitative point of view.

The isotherms corresponding to $T \geqslant T_c$ correctly reproduce the shape of the experimental isotherms for temperatures above or equal to the critical temperature. On the other hand for $T < T_c$ a Van der Waals' isotherm displays a maximum and a minimum in place of the isobaric region of lique-faction observed experimentally.

Thus Van der Waals' equation itself does not account for the existence of the isobaric region. This failure is easy to understand. We have implicitly assumed the existence of a uniform density $n = N/V$. In fact it is precisely in the liquefaction region, where gas and liquid with different densities exist together, that the assumption is incorrect.

To correctly describe the facts, a part of the Van der Waals' isotherm must be replaced *ad hoc* by a horizontal line, as indicated in Fig. 1.9. We

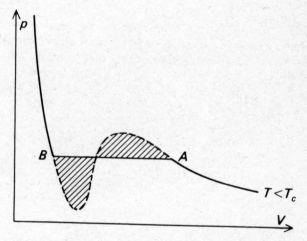

Figure 1.9. The replacement of part of a Van der Waals' isotherm by an isobaric region.

will see later that this line must be drawn such that the shaded areas are equal; this is known as *Maxwell's construction*. In this way, we obtain, if not quantitative agreement, at least a satisfying qualitative description of the real isotherms.

The dashed part of the isotherm represents the equation of state that would result if the fluid remained uniform instead of separating into gas and liquid. The dashed portions near the extremities of the isobaric region of liquefaction can in fact be observed. They represent *metastable* states.

It is possible for example to compress the gas isothermally a little beyond point *A* without it liquefying. A *supersaturated vapour* is then obtained. Such a state is not a state of stable equilibrium; it is said to be *metastable*. A sudden change of conditions for instance will result in condensation and terminate the metastable state. In the same way beyond point *B* it is possible to obtain a metastable liquid, ready to vaporize under the action of some perturbation. These phenomena have technical applications, in the form of the Wilson cloud chamber and the bubble chamber, to the observation of elementary particles.

A Wilson cloud chamber consists of an enclosure containing a vapour. The vapour is expanded suddenly by moving a piston. As we shall see later, this has the effect of cooling the vapour and bringing it into a metastable state. A charged particle that traverses the chamber produces ions along its path and in the vicinity of the ions condensation is induced. In this way the trajectory of the particle can be seen as a line of small droplets.

In the bubble chamber, a more recent development, a liquid such as liquid hydrogen is made metastable by expansion. Bubbles of vapour are then observed to form along the trajectory of a particle.

THE LAW OF CORRESPONDING STATES

We have determined the critical constants of a fluid starting from the parameters *A*, *B*, and *R* of Van der Waals' equation. Inverting these relations, we find

$$A = 3p_cV_c^2; \quad B = \tfrac{1}{3}V_c; \quad R = \frac{8}{3}\frac{p_cV_c}{T_c}$$

By substituting these expressions into Van der Waals' equation the *reduced equation* is obtained

$$\frac{p}{p_c} = \frac{8}{3}\frac{T/T_c}{V/V_c - 1/3} - \frac{3}{(V/V_c)^2}$$

It can be seen that the *reduced variables* p/p_c, T/T_c, and V/V_c are related by an equation of state which now contains only numerical coefficients independent of the nature of the fluid. Two different fluids for which two of the reduced variables, V/V_c and T/T_c for instance, are equal are said to be in *corresponding states*. The other reduced variable, p/p_c in our example, must then be the same for both gases. In other words, any two fluids must have the same family of isotherms whenever the reduced variables are used.

24

We have just established the law of corresponding states starting from Van der Waals' equation. In fact the same reasoning can be reproduced regardless of the equation of state considered, provided it contains only three parameters. These parameters are expressed as functions of V_c, T_c, and p_c and, from dimensional considerations, the equation of state must necessarily take the form of a relation between the reduced variables V/V_c, T/T_c, and p/p_c, independent of the nature of the fluid. For a group of fluids to obey the law of corresponding states, it is sufficient that they are all described by one equation of state containing only three parameters.

If the law of corresponding states is examined, some interesting information can be derived about the intermolecular forces. In principle of course, the equation of state must be deducible from the intermolecular forces, although the actual calculation can be made only with the help of very large computers or by the use of approximations. If the intermolecular potential contains only two parameters that depend on the nature of the fluid, as is the case in the Lennard-Jones potential with parameters ε and σ, these parameters plus Boltzmann's constant k are available to construct an equation of state. Since this equation depends on three parameters, a law of corresponding states can be established. It is evident that for a group of fluids to obey this law it is sufficient that they are all described by an intermolecular potential containing only two parameters.

In fact the law of corresponding states is poorly supported by experiment. Certainly groups of fluids exist, whose members have very similar families of isotherms in reduced coordinates, but the families differ considerably from one group to another. This is evidence that an intermolecular potential depending on only two parameters is inadequate to accurately describe every fluid. However, the law of corresponding states has a general approximate validity which is the reason for its usefulness.

THE CRITICAL REGION

The properties of a fluid in the vicinity of its critical point are difficult to study experimentally, and moreover, present troublesome theoretical problems.

The critical isotherm ($T = T_c$) has a horizontal tangent at the critical point. At that point

$$\frac{\partial p}{\partial V} = 0$$

or taking the numerical density $n = N/V$ as a variable instead of V

$$\frac{\partial p}{\partial n} = 0$$

25

When the density undergoes an infinitesimal variation from the critical point, the pressure varies only by a higher order infinitesimal; in other words, the density is not well determined from the knowledge of the pressure. The result of this is that if a fluid is in equilibrium under critical conditions, its pressure is uniform but its density has a tendency to fluctuate spontaneously from point to point. These fluctuations of density are the cause of the phenomenon of *critical opalescence*. If a sealed glass tube, known as a Natterer tube, contains a mass of fluid such that the volume of the tube is precisely the corresponding critical volume, when the tube is brought to the critical temperature its contents become opalescent. This is because light is strongly scattered by the inhomogeneities in density inside the fluid. In fact, one method of measuring the critical temperature is to note the temperature at which opalescence occurs.

These inhomogeneities in density are the source of difficulty when constructing a well-behaved theory of fluids in the critical region. Simple-minded theories of the type that lead to Van der Waals' equation, apply only if the density is uniform and they are usually found wanting in the region of the critical point.

Let us consider for example the shape of the critical isotherm in the region of the critical point. According to Van der Waals' equation, p is a well-behaved function of V near the critical point and can be expanded as a Taylor series around $V = V_c$. Since $\partial p/\partial V = \partial^2 p/\partial V^2 = 0$ at the critical point, the expansion begins with a term of third order

$$p - p_c = -C(V - V_c)^3 + \dots$$

where C is a constant. Behaviour of this type is not peculiar to Van der Waals' equation, but is found in all simple theories. A careful experimental study of the critical isotherm shows that near the critical point it has the form[13]

$$p - p_c \sim [\mathrm{sgn}\,(V_c - V)]\,|V - V_c|^\delta$$

with an exponent δ of about 4·2. The theoretical justification for this exponent is not yet understood and presents a problem on which physicists are still working.

Following the same train of ideas, let us consider a mixture of gas and liquid that must of course be at a temperature below the critical temperature. Let n_g and n_l be the numerical densities of the gas and the liquid at temperature T (Fig. 1.10). At the critical temperature gas and liquid become one and the same; the difference $n_l - n_g$ tends to zero when T tends to T_c. Simple theories predict a behaviour of the type

$$n_l - n_g \sim C(T_c - T)^{\frac{1}{2}}$$

---

[13] sgn (x) is a function of x with the value $+1$ or -1 according to whether x is positive or negative.

26

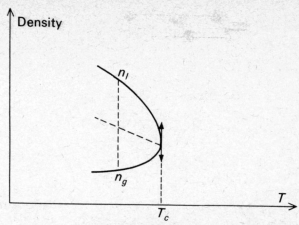

Figure 1.10. The densities n_g and n_l of a gas and a liquid in mutual equilibrium at temperature T. The mid-points of the lines joining n_g and n_l at the same temperature lie almost on a straight line; this is known as the law of rectilinear diameters.

while experiment shows

$$n_l - n_g \sim C(T_c - T)^\beta$$

with an exponent β of about $\frac{1}{3}$.

1.7 Solids

We will complete our study of the states of matter with a brief description of solids.

Crystalline and amorphous solids have distinct properties. The atoms of an amorphous solid, glass for example, are arranged in an irregular fashion, like those of a liquid. In a sense such a substance is more like a viscous liquid than a true solid.

In contrast, the atoms of a crystalline solid are distributed in a geometrically regular arrangement called a lattice. The whole crystal may be generated by translations of a unit cell which is repeated identically. By way of examples, Fig. 1.11 represents the simple cubic lattice of sodium chloride, while Fig. 1.12 represents the diamond lattice. Usually crystalline solids consist of an aggregate of microcrystals. For example, the common metals have this form unless specially prepared.

The binding of the crystal lattice is due to the forces between the atoms. These forces are of various types depending on the solid considered, but nearly always have the same general behaviour as we have discussed already: the atoms attract one another at large distances and repel one another at

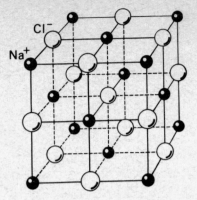

Figure 1.11. The sodium chloride lattice. The chlorine atoms ○ and sodium atoms ● occupy alternate corners of the cubes.

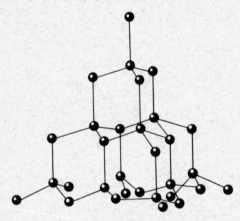

Figure 1.12. The diamond lattice. Each carbon atom is at the centre of a tetrahedron which has four other atoms at its corners.

short distances. The atoms tend to take up equilibrium positions in a perfect crystal lattice. Thermal vibration opposes this tendency and each atom oscillates about its mean position.

1.8 Phase changes of a pure substance

In general a substance can exist in different physical states such as gas, liquid, or solid. These states are sometimes referred to as the gaseous phase, the liquid phase, or the solid phase respectively.

28

Changes of phase satisfy particularly simple laws when a pure substance, as distinct from a mixture, is concerned and we will restrict ourselves to this case. We will also assume that the substance does not decompose in the temperature range under consideration.

When heated, a solid, for example ice, frequently behaves in the following way. At a particular temperature the solid melts. If it is a 'true' solid, i.e., crystalline, and we will limit ourselves to that case, the melting is sharp. At this well-defined temperature, the solid is converted sharply into a liquid. As long as the melting is incomplete, the temperature remains constant. If the heating is continued, the liquid reaches a higher temperature at which it boils and is converted into a gas. Once again the temperature stays constant as long as the boiling continues. When the boiling is complete, the gas can be heated to higher temperatures.

It can also happen that the heated solid changes directly to the gaseous state, for example solid carbon dioxide in air. Such solids are said to *sublime*.

On cooling once more the inverse phenomena are observed at the same temperatures.

The study of phase changes in free air is complicated by the presence of the air and by the fact that any gas formed escapes; a state of equilibrium is not achieved. Therefore to begin with we will study the simplest example in which the substance under consideration is confined in an enclosure on its own. Any two of the three variables p, V, and T can be varied, since the third is a function of the other two, and the behaviour of the system followed.

LIQUID–GAS PHASE CHANGE

We have already prepared the ground for the study of phase changes by discussing real fluids. In this connection we made use of a graphical representation in the (p, V) plane. An important result of this study was the existence of an isobaric region of liquefaction where all points represented possible states of equilibrium between the liquid and gaseous phases (Fig. 1.13).

While, normally, knowledge of the pressure and the temperature determines the volume, in this isobaric region (which corresponds to a particular pressure and temperature) the system can take an infinite number of states with different volumes, ranging from a pure liquid with volume V_l through all the states in which gas and liquid coexist in all possible proportions to a pure gas with volume V_g. For a fluid of total mass M, liquid of mass xM and gas of mass $(1 - x)M$ coexist in a state of volume V where

$$V = xV_l + (1 - x)V_g \quad (0 \leqslant x \leqslant 1)$$

Rearranging this equation, we find

$$\frac{V_g - V}{V - V_l} = \frac{x}{1 - x}$$

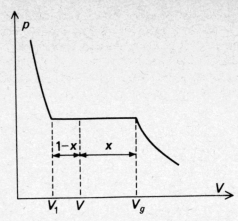

Figure 1.13. The isobaric region of liquefaction represents the states of coexistence of liquid and gas. $(V_g - V)/(V - V_l)$ is equal to the ratio of the masses of liquid and gas $x/(1 - x)$.

so that the point representing this state of coexistence divides the horizontal part of the isotherm in the ratio of the masses of liquid and gas.

The states may be represented in a different way in the (p, T) plane. The pressure p in the isobaric region, where liquid and gas are in equilibrium, is a function of the temperature T. This pressure is called the *saturated vapour pressure*, because the vapour is said to be saturated when it is in equilibrium with the liquid (on the other hand, the vapour is called *unsaturated* when the system consists of the gaseous phase alone). The saturated vapour pressure as a function of temperature is shown in Fig. 1.14. The points on the curve correspond to the states of equilibrium between liquid and

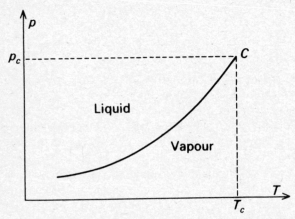

Figure 1.14. The liquid–vapour equilibrium in the (p, T) plane. C is the critical point.

vapour. At a higher pressure or a lower temperature the system consists only of liquid, whereas at a lower pressure or a higher temperature it consists only of gas. The equilibrium curve separates these regions of the (p, T) plane, which represent the pure liquid states and the pure gaseous states respectively. The point C at one end of the curve is the *critical point*. Beyond that, liquid and gas become indistinguishable and it is possible to go continuously from one to the other.

Evaporation. If a liquid is introduced into an enclosure which has previously been evacuated, some of the liquid vaporizes almost immediately so that the gas pressure becomes the saturated vapour pressure. If there is not enough liquid for this to happen all the liquid vaporizes and there is nothing but unsaturated vapour.

If the enclosure contains a gas such as air above the liquid, the liquid evaporates slowly instead of suddenly. However, the equilibrium eventually reached is not modified much by the presence of the air. To a first approximation, the total pressure in equilibrium is the sum of the initial air pressure and the saturated vapour pressure of the pure liquid–gas system.

If a liquid has a surface open to the atmosphere, the resulting vapour escapes, the saturated vapour pressure is never reached, and the liquid evaporates until there is none remaining.

This evaporation arises from a net balance between the molecules which are evaporating and those which are condensing. The rate of evaporation is therefore increased by removing the vapour as soon as it is formed, by placing the liquid in a current of air for instance.

The molecules which evaporate are those which have sufficient kinetic energy to overcome the attraction of the other molecules in the liquid; these of course have the greatest velocity. The molecules which stay in the liquid have a lower mean velocity and as a result of this the liquid tends to cool while it is evaporating. Condensation, on the other hand, tends to produce heating.

Boiling. The rate of evaporation increases with temperature since, when the molecules of the liquid are in more vigorous thermal motion, they escape more easily. When the temperature reaches a certain value, the effect changes in character. Instead of being formed at the surface, the gas is generated within the bulk of the liquid in the form of bubbles which come bursting to the surface; the liquid is said to boil.

For a given pressure, a liquid boils at a well-defined temperature which remains constant throughout the boiling. *The boiling point is the temperature at which the saturated vapour pressure is equal to the pressure to which the liquid is subjected*; in other words, the plot of p against T for equilibrium between liquid and vapour also represents the relation between the pressure and the temperature at boiling point. In fact the gas bubbles can only exist within the liquid in conditions of equilibrium between liquid and vapour; at a lower temperature the interior of the liquid is stable, while at higher temperatures there is only gas.

In a closed vessel, liquid with air above it will not boil. Effectively, the total pressure experienced by the liquid is the saturated vapour pressure plus the air pressure; this total pressure can never be equal to the saturated vapour pressure. In this way, water can be heated in a pressure cooker to temperatures greater than 100°C without boiling. Boiling will be initiated by allowing the steam to escape through a valve; by adjusting the pressure to a value above normal atmospheric pressure with the help of this valve, boiling can be made to occur at a temperature above 100°C.

SOLID–LIQUID PHASE CHANGE

Equilibrium between a crystalline solid and a liquid has a number of points in common with the liquid–gas equilibrium.

In the (p, V) plane, the isotherms still display an isobaric region, which is now the isobaric region corresponding to solidification. The points in the region represent the equilibrium states between solid and liquid.

The phase change is accompanied by a change in volume, which is however not very important since solids and liquids have similar densities. Usually the solid is the denser, as it is in the case illustrated in Fig. 1.15. There are however some notable exceptions; in particular, water is more dense than ice. In this case the labels solid and liquid on Fig. 1.15 must be reversed. The branches of the isotherm beyond the isobaric region have

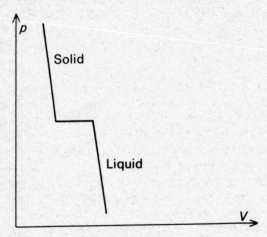

Figure 1.15. The isobaric region of solidification in the (p, V) plane.

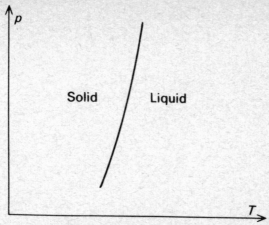

Figure 1.16. The solid–liquid equilibrium curve for a normal substance.

been drawn with very steep slopes because liquid and solid are both almost incompressible.

The phase change can also be represented in the (p, T) plane. A curve $p(T)$ can be drawn which gives the equilibrium pressure as a function of the temperature and separates the liquid and solid regions. In the usual case, where the solid is more dense than the liquid, a rise in pressure at the equilibrium temperature promotes the formation of the solid phase; the slope of the $p(T)$ curve is as drawn in Fig. 1.16. On the other hand, in the exceptional cases where the liquid is more dense (water), the $p(T)$ curve slopes in the opposite direction as in Fig. 1.17. In all cases the absolute value of the slope is very large, because the temperature of the solid–liquid transition is almost independent of the pressure.

A notable distinction between the liquid–gas and the solid–liquid transition is that there is no critical point for the latter, the $p(T)$ curve extending indefinitely towards high pressures. This is accounted for by the fact that, unlike liquid and gas, solid and liquid can never become indistinguishable. The crystalline structure of a solid endows it with an ordered character which distinguishes it absolutely from a fluid and it is impossible to pass continuously from one to the other.

The microscopic mechanism of fusion is qualitatively very simple. As the temperature of a solid rises, the vibrations of the atoms about their mean equilibrium positions increases proportionally. At the melting point, the amplitude of these vibrations has become sufficient to cause the crystalline structure to break up. It is remarkable that for all solids at their melting points, the amplitude of the atomic vibrations is always of the same order of magnitude, namely about 10 per cent of the interatomic distance.

33

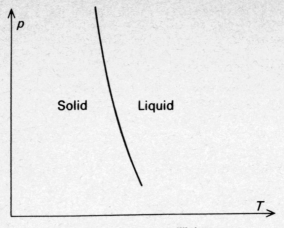

Figure 1.17. The ice–water equilibrium curve.

SOLID–GAS PHASE CHANGE

At a sufficiently low temperature and pressure a solid, camphor for example, *may* sublime, i.e., transform directly to the gaseous state.

The solid–gas equilibrium has a strong resemblance to the preceding examples; in the (p, V) plane the isotherms display an isobaric region. The sublimation pressure is a function $p(T)$ of the temperature as in Fig. 1.18, where the typical curve separates a solid region in the (p, T) plane from a gaseous one. There is no critical point because we are concerned with a solid–fluid equilibrium.

Strictly speaking any solid will always have a tendency to sublime if the pressure on the surface is not maintained the same as the sublimation

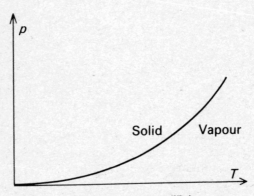

Figure 1.18. The solid–vapour equilibrium curve.

pressure. In fact sublimation usually takes place at such a low rate that it is unobservable. Everyday solid objects with which we are familiar do not sublime appreciably!

PHASE DIAGRAM. TRIPLE POINT

Let us plot the various equilibrium curves that we have considered on the same graph in the (p, T) plane.

For very many substances a diagram of the type shown in Fig. 1.19 is obtained. The different equilibrium curves meet at point t, which is known as the triple point. Under the unique temperature and pressure conditions at point t, the three phases are simultaneously in equilibrium with one another.

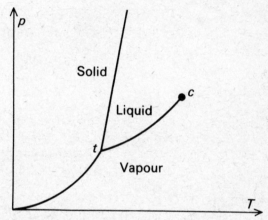

Figure 1.19. Different phases in the (p, T) plane. C is the critical point and t the triple point.

The full isotherms in the (p, V) plane have the form shown in Fig. 1.20. At a temperature below the triple point temperature T_t, the gas changes to a solid when compressed. Between T_t and the critical temperature T_c, compressing the gas yields first the liquid then the solid. Finally, above the critical temperature, there is no distinction between liquid and gas and the fluid, as this state of the substance can be called, becomes a solid when compressed.

Certain substances have different types of phase diagrams. If the solid–liquid equilibrium curve has a negative slope, as for example in the case of water, the appearance of the isotherms is modified in comparison with plots of Fig. 1.20. There can be several solid phases distinguished, for example, by different crystalline structures and these phases would be separated in the

35

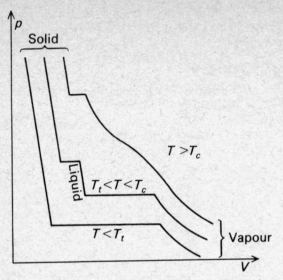

Figure 1.20. The isotherms in the (p, V) plane for a normal substance.

(p, T) plane by additional equilibrium lines which would further complicate the graph. Helium, in both isotopic forms 3 and 4, has a unique phase diagram.

SURFACE OF STATE

The two-dimensional graphs of Figs. 1.19 and 1.20 provide a partial representation of the function $p(V, T)$, one of the variables being held constant. The complete representation of the equation of state $f(p, V, T) = 0$ must be plotted in three-dimensional space. The coordinate axes are p, V, and T and the equation of state describes a surface. This surface of state is shown in Fig. 1.21 for a typical substance. The two-dimensional plots (p, V) or (p, T) are the projections of this figure on the corresponding planes.

Problems

1.1 Equation of state. The equation of state for 10^{-3} kg of a gas can be written

$$\left(p + \frac{185 \times 10^{-6}}{V^2}\right)(V - 5 \times 10^{-6}) = 2.2\, T$$

where p, V, and T are expressed in SI units, i.e., Nm^{-2}, m^3 and K.

Under normal pressure and temperature conditions, calculate

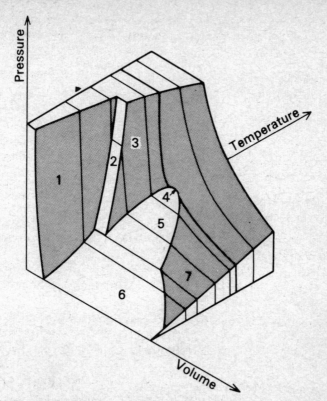

Figure 1.21. The surface of state for a normal substance. (1 = solid, 2 = solid–liquid, 3 = liquid, 4 = critical point, 5 = liquid–vapour, 6 = solid–vapour, 7 = vapour).

(a) the volume of the gas

(b) the isothermal coefficient of compressibility

$$\chi = -\frac{1}{V}\left(\frac{\partial V}{\partial p}\right)_T$$

(c) the isobaric coefficient of expansion

$$\alpha = \frac{1}{V}\left(\frac{\partial V}{\partial T}\right)_P$$

(d) the isometric pressure coefficient

$$\beta = \frac{1}{p}\left(\frac{\partial p}{\partial T}\right)_V$$

1.2 Show that independent of the equation of state chosen to describe the gas there exists a simple relation between the three coefficients χ, β, and α defined in problem 1.1.

1.3 Experiments have shown that for a real gas, the coefficients defined in 1.1 can be written

$$\alpha = \frac{A}{AT + Bp}; \qquad \chi = \frac{1}{p} - \frac{B}{AT + Bp}$$

where A and B are constants. Deduce the equation of state and compare with Van der Waals' equation.

1.4 The gas whose equation of state is given in problem 1.1 is chemically pure. Can you identify it?

1.5 Derive expressions for the critical coordinates of a gas which obeys an equation of state of the form

$$p(V - b)\exp\left(\frac{a}{VRT}\right) = RT \quad \text{(Dieterici's equation)}$$

1.6 The equation of state of 10^{-3} kg of a gas can be written as

$$\left(p + \frac{a}{V^2}\right)(V - b) = rT$$

where a, b, and r have the same numerical value as in problem 1.1. Derive expressions for the functions $A(T)$, $B(T)$, and $C(T)$ such that the power series

$$pV = A(T) + pB(T) + p^2C(T)$$

represents an alternative to this equation of state. In what region of the variables p, V, and T is the power series valid?
From these expressions, determine the temperature T_1 at which

$$\left(\frac{\partial pV}{\partial p}\right)_{p=0} = 0$$

and the temperature T_2 at which

$$\left(\frac{\partial^2 pV}{\partial p^2}\right)_{p=0} = 0$$

Sketch the isotherms at the temperatures T_1, T_2, and T_c in Amagat's representation.
A constant volume thermometer with a volume of 5×10^{-3} m^3, contains 10^{-3} kg of this gas. The pressures measured at the temperatures of melting

38

ice and of boiling water under atmospheric pressure determine the points 0 and 100 on the scale of temperature of this thermometer. At a temperature of 30 on this scale, what is the temperature in degrees Celsius if the gas obeys (a) the original equation of state and (b) the power series $pV = f(p)$?

1.7 Hydrogen is compressed to 200 atmospheres at 20°C in a cylinder whose volume is $10^{-1}\,m^3$. Assuming that hydrogen behaves like a perfect gas, calculate

(a) the number of moles of gas in the cylinder
(b) the mass of gas
(c) the pressure expressed both in atmospheres and in newtons per square metre if the temperature is raised to 500°C.

1.8 A balloon with a mass of 300 kg when empty is filled with helium at atmospheric pressure and 20°C. Assuming helium to be a perfect gas, calculate the mass of helium required if the balloon is to begin to rise. What is the corresponding volume?

1.9 Extend the relation $p = \frac{1}{3}(N/V)m\overline{v^2}$ to the case of molecules striking an irregular wall. Assume the distribution of velocities to be unchanged by collisions with the walls.

1.10 What is the root mean square velocity $\sqrt{\overline{v^2}}$ of oxygen molecules at 30°C (make use of the relation $p = \frac{1}{3}(N/V)m\overline{v^2}$).

1.11 Correction to the fundamental interval. A mercury thermometer with a linear scale reads -3 in melting ice and $+105$ in boiling water at atmospheric pressure. What is the temperature when it reads $+25$? At what temperature is this correction to the fundamental interval zero?

1.12 Correction for the emergent stem. A mercury thermometer is partly immersed in a bath at temperature T. When the stem is dipped in as far as the mark corresponding to 15° it reads 50°, while if the stem is dipped in as far as 40° it reads 50·18°. If the outside temperature is 15°C, what is the temperature T?

1.13 A platinum resistance thermometer reads 10·0000 ohms in melting ice, 13·9194 ohms in boiling water at atmospheric pressure, and 26·3716 ohms in boiling sulphur at atmospheric pressure. From these figures deduce a numerical relation expressing the resistance as a function of temperature defined by the international scale. What is the temperature corresponding to a resistance of 16·3000 ohms?

1.14 A pressure cooker of volume $10^{-2}\,m^3$ is filled with a mass of water, m, at a temperature of 20°C and atmospheric pressure. If it is closed and its temperature raised to 120°C, what is the pressure if m is (a) $6 \times 10^{-3}\,kg$

and (b) 16×10^{-3} kg? Assume that the unsaturated water vapour behaves like a perfect gas and that the saturated vapour pressure of water is $(t/100)^4$ atmospheres, where t is the temperature in degrees Celsius.

1.15 What is the atmospheric pressure at the summit of Mont Blanc (height 4 807 m) if the atmosphere is assumed to have a uniform temperature of 0°C. At what temperature does water boil there?

1.16 A vertical tube of uniform cross-section $s = 10^{-4}$ m², closed at its lower end, contains a certain mass of air enclosed by a length of mercury.

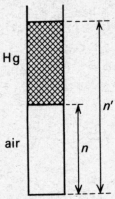

The mercury surfaces are $n = 0.5$ m and $n' = 0.6$ m from the bottom of the tube. What are the new values of n and n' if the tube is (a) inverted and (b) made horizontal? Take the conditions outside the tube as 'normal', i.e., pressure = 1 atmosphere and temperature = 0°C.

1.17 Consider the arrangement of problem 1.16 again. If 10^{-6} m³ of ether is introduced into the air enclosed by the length of mercury, what are the values of n and n' in the two positions of the tube? The vapour pressure of ether at 0°C is 0·100 m Hg.

1.18 A vertical U-tube of cross-section 10^{-4} m² has a short closed arm and a long arm that is open to atmospheric pressure. The closed arm is full of

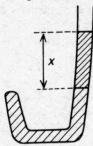

mercury and the long one filled with mercury to a level $x = 0.05$ m above the level in the closed arm. If a small mass of water, of negligible volume when liquid, is introduced into the short arm and the tube and its contents heated to 100°C, what is now the height x of the mercury in the long arm?

1.19 In the preceding problem, a small amount of air 60×10^{-9} kg is introduced as well as the mass of water. What is the new height x of the mercury?

2. Internal energy and the first law of thermodynamics

2.1 Internal energy

Let us consider a macroscopic system, i.e., any collection of bodies.

On an atomic scale, this system consists of particles which are endowed with kinetic energy and, in general, mutual potential energy representing the interaction of the particles among themselves. The energy associated with the internal mechanics of the system on a microscopic scale is called the *internal energy*.

For example in the case of a monatomic perfect gas, the internal structure of the molecules makes no contribution to the internal energy and, since the molecules do not interact with one another, they have no mutual potential energy. The internal energy U reduces to the kinetic energy of translation of the molecules

$$U = U_t = \sum \tfrac{1}{2} m v_i^2$$

where v_i is the velocity of the ith molecule and the summation is taken over all the molecules.

Like all mechanical energy, the internal energy can only be determined to within an arbitrary additive constant. If required, an absolute definition of internal energy can be given by means of Einstein's formula. If M is the mass of the system, assumed at rest on a macroscopic scale, and c is the velocity of light, the internal energy is simply Mc^2. This absolute definition plays no part in ordinary thermodynamic phenomena.

The internal energy U of a system in equilibrium is a constant which is a function of the macroscopic state of the system. U is called a *state function*. For a fluid, the macroscopic state is determined by two of the three variables (p, V, T) and U can be considered as a function of the two chosen independent variables. Alternatively, it is possible and often convenient to take U and V as the independent variables.

The internal energy of a system varies when the system exchanges energy with its surroundings. The energy exchanged by means of the random interaction of molecules of the system with those of the surroundings is called *heat*. The energy exchanged by the action of external forces that are directed on a macroscopic scale is called *work*; for example, energy is exchanged in the form of work, when the volume of the system is varied.

Since internal energy, work, and heat are all forms of energy, they are measured in the same units, namely joules in the SI system.

2.2 Reversible and irreversible transformations

For the remainder of this book it will be important to distinguish between reversible and irreversible transformations.

For a system in a state of equilibrium, certain macroscopic parameters, such as volume, pressure, and temperature, have well-defined values. On the other hand when the system is not in equilibrium, these parameters are often poorly specified; for example, pressure and temperature can differ from point to point and thus not have a unique well-determined value. For this reason systems in equilibrium are particularly simple to describe very precisely.

When the state of a system changes, the system is said to undergo a *transformation*. For example, the volume of the enclosure containing the system can be varied, the system can be heated and so on. In general these operations will perturb the state of the system which, temporarily at least, will be displaced from equilibrium.

However under certain special conditions the system can be transformed in such a way that at every instant it is almost in a state of equilibrium. To ensure this, changes of volume must be made very slowly to avoid the creation of currents of matter, inhomogeneities in pressure and so on. For changes of temperature the system must be placed in surroundings whose temperature is very little different from its own. These surroundings are called a *source of heat* whether heat is supplied or absorbed.

For such a transformation, very small changes in the external parameters are sufficient to reverse its direction. If the volume is increasing, a very small increase in the external pressure on the enclosure walls is enough to cause the volume to decrease. If the system is in the process of absorbing heat from a source slightly hotter than itself, lowering the temperature of the source a little is sufficient to cause the system to lose heat. This is why such transformations are called *reversible*.

In brief, a reversible transformation is such that an infinitesimal change in the external conditions suffices to reverse the direction of the transformation. The state of the system remains infinitely close to a state of equilibrium throughout the transformation. In consequence, the transformation occurs infinitely slowly.

A reversible transformation is an ideal limiting case which can only be realized approximately in practice. A transformation which is not reversible is said to be *irreversible*. Transformations which occur spontaneously in nature are irreversible. The expansion of a gas from one container into another is an irreversible phenomenon since the gas is not in equilibrium while it is flowing.

A phenomenon involving the dissipation of energy, such as the passage of an electric current through a resistance, is essentially irreversible; regardless of the direction of the current, heat is always produced, never absorbed. In passing, it should be noted that this irreversibility is bound up with the fact that the flow of electricity does not take place infinitely slowly. The energy released by Joule heating in a resistance R due to a current I flowing for time t is

$$RI^2t = \frac{Rq^2}{t}$$

where $q = It$ is the total charge that has passed through the resistance. It can be seen that for a given charge q, the energy dissipated by the irreversible Joule heating tends to zero as the time for which the electricity flows tends to infinity.

2.3 Work

The amount of energy exchanged in the form of work between a system and its surroundings will be denoted by W. We will make the convention that W is positive if the work is done on the system and negative if the work is done by the system.

An infinitesimal quantity of work will be denoted by δW. We will reserve the symbol d to indicate the difference between two infinitely close values of one of the variables characterizing the system such as in dp, dV, dT, dU and so on. The reason for writing δW rather than dW is to emphasize that δW is not a difference between two neighbouring values of a work variable. The word 'work' applies to the energy while it is being transferred, but work is not a variable that can be associated with a given state of the system.

Work appears in various ways. For example, when an electric current passes through a system, electrical work is done. If an external source sets up a potential difference V across the system and feeds into it a current I for a time dt, the work done on the system is

$$\delta W = VI\,\mathrm{d}t$$

At the end of a finite time, the work done is

$$W = \int VI\,dt$$

We will be particularly interested here in mechanical work done by the forces responsible for the pressure on a system, since energy may be given to the system by compressing it. We will calculate the work δW done on the system when its volume changes by dV.

Let us consider (Fig. 2.1) an element of wall of surface area dS, which has been displaced normally by dx, where we have taken the positive direction

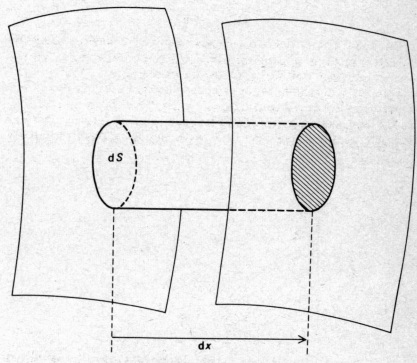

Figure 2.1. Change of volume of a system.

outwards. Let the pressure exerted by the external forces on the element of the walls be p_0. The force corresponding to this pressure, which is directed inwards, is $-p_0\,dS$ and it does work

$$-p_0\,dS\,dx = -p_0\,d\tau$$

where $d\tau$ is the volume of the small cylinder swept out by dS. Assuming that the pressure p_0 due to the external forces is the same over the whole surface, the work done on the system is

$$\delta W = -p_0 \, dV$$

where dV is the increase in the total volume.

This formula, with p_0 representing the external pressure, can be applied very generally; it is not necessary to assume that the transformation is reversible.

In the particular example of a system characterized by a well-defined internal pressure p that is infinitely close to the external pressure p_0, we may use the *internal* pressure to calculate δW and write

$$\delta W = -p \, dV$$

This holds for a fluid which undergoes a reversible transformation; the formula is particularly useful then, because p can be expressed in terms of the other variables with the help of the equation of state.

There is a useful graphical way of representing the work done on a system. A macroscopic state of a fluid in equilibrium is defined by the two variables p and V and can thus be represented by a point in the (p, V) plane. During a reversible transformation,[1] the fluid passes through a series of states of

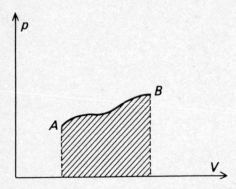

Figure 2.2. The representation in the (p, V) plane of a reversible transformation and the work done during such a transformation.

equilibrium each of which can be represented by a point; these points describe a curve in the (p, V) plane (Fig. 2.2). During the transformation from the initial state A to the final state B, the total work done on the fluid, i.e.,

[1] In general during an irreversible transformation, the fluid pressure p does not have a well-defined value and it is no longer possible to represent the transformation by a curve.

the sum of all the infinitesimal amounts of work $-p\,dV$, is the integral

$$W = - \int_A^B p\,dV$$

$-W$ is equal to the shaded area in Fig. 2.2.

A transformation which brings the system back to its initial state is said to be *cyclic*. A reversible cyclic transformation is represented by a closed curve in the (p, V) plane (Fig. 2.3). The work done on the system is

$$W = - \oint p\,dV$$

where the symbol $\oint$ denotes that the integral is taken along a closed curve. If the system follows the curve in the direction shown in Fig. 2.3 it is easily seen that the work done on the system is equal to the area enclosed by the curve. If the system follows the curve in the opposite sense, the work done on the system is negative but has the same magnitude as before.

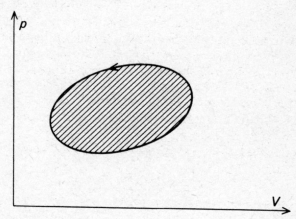

Figure 2.3. The representation in the (p, V) plane of a reversible cyclic transformation and the work done during such a transformation.

2.4 Heat

The amount of energy exchanged in the form of heat between a system and its surroundings will be denoted by Q. As in the case of work we will make the convention that Q is positive if the heat is absorbed by the system and negative if it is supplied by the system. For an infinitesimal amount of heat, we will use the notation δQ; we choose to write δ rather than d for the same reasons given in the discussion about work.

Transfer of energy in the form of heat takes place when the molecules of the system interact in a *disordered* way with the molecules of the surroundings. This happens when a cold body is placed in contact with a hot one; the more agitated molecules of the hot body strike the molecules of the cold one and on the average impart to them some of their energy. This process is fundamentally different from the transfer of energy in the form of work, such as occurs for example during the compression of a fluid by means of a piston. In the final analysis it is true that even in this case the molecules of the piston interact with those of the fluid. The difference is that when the piston is displaced all its molecules have an ordered movement superimposed on their thermal motion and it is the contribution of this ordered movement to the transfer of energy that is called work.

Heat and temperature must not be confused. Heat is a form of energy and is an extensive quantity since in order to produce a given change in the state of a substance a quantity of heat proportional to the mass must be supplied. Temperature, on the other hand, is an intensive quantity which characterizes the state of a substance independently of its mass.

In fact it can happen that heat is transferred to a system without changing its temperature. This occurs when water boils for example. While the water is boiling the temperature remains at 100°C and the heat transferred serves only to increase the mutual potential energy of the molecules. This energy must be supplied to separate the water molecules, which are bound to one another, and thus produce steam.

Conversely a temperature variation may be induced in a system without supplying it with heat. For example when a gas is compressed, its temperature rises.

Generally however when heat is transferred to a system, its temperature changes.

2.5 The first law of thermodynamics

The first law of thermodynamics is merely a statement of the conservation of energy: **any change in the energy of a system is equal to the energy received by it during the change.**

In thermodynamics the systems that are studied have internal energies that vary by means of exchange of work and heat with the surroundings. If the internal energy changes from the value U_1 to the value U_2 and if W and Q are respectively the work and the heat absorbed by the system during the transformation[2] we have

$$U_2 - U_1 = W + Q$$

[2] It should perhaps be emphasized that the conservation of energy, and therefore the first law, have a general validity and apply to all transformations, reversible and irreversible.

This equation is one form of the first law. For an infinitesimal transformation we can write

$$dU = \delta W + \delta Q$$

The internal energy is a state function; knowledge of the initial and final states determines the change in internal energy $U_2 - U_1$ or dU in the case of an infinitesimal change. On the other hand, as we have already mentioned, work and heat are not state functions; the work and the heat absorbed during a transformation do not depend on the initial and final states alone but on the manner in which the transformation is effected.

For example when 1 g of water at 0°C and atmospheric pressure is taken to 100°C at atmospheric pressure, its internal energy increases by 418 J. However, this transformation may be achieved in different ways. The 418 J can be supplied in the form of mechanical work by stirring the water with a paddle or in the form of heat by putting the water over a flame. In both cases $W + Q = 418$ J, but the individual values of W and Q depend on the method employed.

We cannot therefore speak of the work or the heat contained in a system. Unlike internal energy, work and heat are not variables with particular values corresponding to a given state of the system.

As U is a state function, the increment dU can be obtained by differentiating U. If p and T for example are chosen as independent variables to describe a fluid, the function $U(p, T)$ has for a differential[3]

$$dU = \left(\frac{\partial U}{\partial p}\right)_T dp + \left(\frac{\partial U}{\partial T}\right)_p dT$$

To avoid possible confusion the symbol of the variable that stays constant during the differentiation is written as a subscript after each partial derivative. dU is a differential in the etymological sense of the term, i.e., the difference between two neighbouring values of U. To be precise, dU is an exact differential. On the other hand, δW and δQ are not exact differentials since W and Q are not state functions; a partial derivative such as $(\partial W/\partial p)_T$ is meaningless.

Nowadays, since we know that heat represents a transfer of energy and we understand its nature on an atomic scale, the first law of thermodynamics is almost self-evident. In the course of the development of physics however, the first law had first to be established empirically. It was noticed that the same change of state could be produced by supplying a system with work or heat or both and this led to the proposal that heat and work were equivalent.

[3] A mathematical note on partial derivatives will be found in Appendix V.

Before this equivalence was recognized, different units were used for heat and work. A special unit known as the calorie,[4] is still sometimes employed to measure heat. The calorie is that quantity of heat which must be supplied to 1 g of water to raise its temperature from 14·5 to 15·5°C at atmospheric pressure. The kilocalorie (10^3 calories) is also used. According to recent measurements the calorie has the value $J = 4·1868$ J; this energy J is called the *mechanical equivalent of heat*.

J may be determined by measuring the work needed to produce the required increase in the temperature of a known mass of water. For example we can heat the water by stirring it with paddles and measure the mechanical work expended in the process. More conveniently we can heat it by means of an electrical resistance and measure the electrical work supplied during the heating.

To avoid having to review the value of J in the light of new measurements, the thermochemical calorie has been introduced and *defined* as equal to 4·184 J.

If we denote quantities of heat expressed in calories by Q' and those in joules by Q then we have $Q = JQ'$. To express heat in calories it is necessary to replace Q by JQ' in all equations. In this book we will express all energies including heat in joules and J will never appear.

2.6 Specific heats

Let us consider a system which absorbs a quantity of heat Q in a well-specified manner while its temperature goes from T_1 to T_2. The relation

$$C = \frac{Q}{T_1 - T_2}$$

defines the *mean thermal capacity* in the interval (T_1, T_2). Like the heat Q, the thermal capacity C is an extensive quantity. An intensive quantity characteristic of a substance is obtained by dividing C by the mass of the substance considered. The quotient $c = C/M$ is called the *specific heat per unit mass*. The thermal capacity per mole or molar specific heat can also be used.

The quantity of heat Q and therefore the thermal capacity and the specific heat depend on the conditions under which the transformation from T_1 to T_2 takes place. The initial and final states, which will be assumed to be states of equilibrium so that their temperatures have well-defined values, are not determined by the temperature variable alone; in the case of a fluid for example, the initial and final values of another variable, such as the pressure or the volume, must be specified. Moreover Q and therefore C, depend not

---

[4] Translator's note. The calorie is not a part of the SI system of units.

only on the initial and final states but also on the way in which the transformation is performed; in this connection the example of the raising of the temperature of water by heating ($Q \neq 0$) or by stirring ($Q = 0$) should be recalled.

Thus we must clearly specify the conditions for the transformation if Q and C are to have well-defined values. For given initial and final equilibrium states, the change in the internal energy $U_2 - U_1$ is well-determined. According to the first law

$$Q = U_2 - U_1 - W$$

so that if the working conditions are such that the work done, W, has a definite value, Q will be determined.

For a given substance, two particular types of transformation and the corresponding specific heats play important roles:

(a) The substance is heated and its volume kept constant so that no work is done. The heat absorbed is

$$Q = U_2 - U_1$$

and depends only on the initial and final states. The corresponding specific heat is called the mean specific heat at constant volume in the interval (T_1, T_2).

(b) The substance is allowed to expand freely at constant external pressure p_0 during the heating. The work done is

$$W = -p_0(V_2 - V_1)$$

and is determined by the volumes V_1 and V_2 of the initial and final states. The heat absorbed

$$Q = U_2 - U_1 - W = U_2 - U_1 + p_0(V_2 - V_1)$$

is also well-determined. The corresponding specific heat is called the mean specific heat at constant pressure in the interval (T_1, T_2). It should be noticed that the internal pressure may not be known during the transformation if the transformation is brought about in an irreversible manner, but in the final and initial states this internal pressure has a value p equal to the constant external pressure p_0. Thus we can also write

$$Q = U_2 - U_1 + p(V_2 - V_1)$$

where p is the internal pressure of the substance in the initial and final states.

For each of the transformations (at constant volume or constant external pressure) that we have just considered, the amount of heat absorbed has a well-defined value which does not depend on the reversible or irreversible nature of the transformation. However, it should be remarked in passing that this is not generally the case; if the pressure and the volume both vary during the transformation, only in the particular case of a reversible transformation will the work done be given by

$$W = - \int_1^2 p \, dV$$

and the heat absorbed by

$$Q = U_2 - U_1 + \int_1^2 p \, dV$$

where p is the internal pressure.

The *true* thermal capacities and specific heats at a given temperature as distinct from the mean values over an interval of temperature are defined by considering an infinitesimal rise in temperature dT produced by an amount of heat δQ

$$C = \frac{\delta Q}{dT}, \qquad c = \frac{1}{M} \frac{\delta Q}{dT}$$

and taking the limit as dT tends to zero. Once again, the conditions of the transformation must be specified. As before we will consider the two cases of particular importance.

(a) When the volume is kept constant, $\delta Q = dU$. The *specific heat at constant volume* is

$$c_V = \frac{1}{M} \left(\frac{\delta Q}{dT} \right)_V = \frac{1}{M} \left(\frac{\partial U}{\partial T} \right)_V$$

(b) When the pressure is kept constant

$$\delta Q = dU + p \, dV$$

The *specific heat at constant pressure* is

$$c_p = \frac{1}{M} \left(\frac{\delta Q}{dT} \right)_p = \frac{1}{M} \left[\left(\frac{\partial U}{\partial T} \right)_p + p \left(\frac{\partial V}{\partial T} \right)_p \right]$$

In the case of solids and liquids, the specific heat at constant pressure is usually considered because these substances are allowed to expand freely at atmospheric pressure during measurements.

A true specific heat depends in general on the state of the system, i.e., on two independent variables such as pressure and temperature. In the first approximation, this dependence can be neglected in certain cases. For example the specific heats of a solid or liquid can be taken as constant in the region of atmospheric pressure and ordinary temperatures. The specific heat of water is constant to almost 1 per cent throughout the interval from 0 to 100°C; according to the definition of the calorie, the specific heat has the value 1 calorie per gram per degree, i.e., 4·18 J per gram per degree.

2.7 Calorimetry

Calorimetry is concerned with the measurement of the quantity of heat. The measurements are performed with a calorimeter which is essentially an adiabatic container, i.e., one that is impermeable to heat and encloses the system under investigation. Such a container can be realized by suspending the system in a vacuum or by surrounding it with a wall that is maintained at the same temperature as the system. Better still, a combination of these two techniques can be used.

The measurements are always based on the principle of the conservation of energy.

MEASUREMENTS BASED ON A BALANCE BETWEEN HEAT AND WORK

We wish to supply a body with a known quantity of heat. To do this we can heat it by means of an electrical resistance. If the thermal capacity of the resistance is neglected the body absorbs a quantity of heat Q equal to the electrical work received by the resistance. Thus if we measure the electric current I in the resistance and the potential difference V across it during the time t, we have

$$Q = \int VI\,dt$$

We can now determine the internal energy of the body to within an additive constant and its thermal capacity. If U_0 is the initial, unknown internal energy, the internal energy after the body has absorbed an amount of heat Q is $U = U_0 + Q$.[5] Measurements of the temperature T of the body are taken during the experiment with, for example, a thermocouple, and the internal energy as a function of temperature, $U(T)$, is obtained.

[5] We are assuming here that the body is not doing any work. If the body expands by ΔV under an external pressure p_0, we should really write

$$U = U_0 + Q - p_0\Delta V$$

In fact for a solid or liquid under atmospheric pressure the work $-p_0\,\Delta V$ is negligible. If the body is placed in a vacuum as is effectively the case in certain types of calorimeter, this work is in fact zero.

The thermal capacity at the temperature T is the derivative

$$C = \frac{dU}{dT}$$

Usually the body is allowed to expand freely at constant pressure and the thermal capacity thus obtained is C_p.

MEASUREMENTS BASED ON THE EXCHANGE OF HEAT

If two systems A and B are brought together and exchange heat until thermal equilibrium is reached, we know from our previous discussions that the amount of heat absorbed by body B can be deduced from the modifications observed in its state. This quantity of heat is equal to that given up by body A, which can therefore be determined.

Let us suppose for example that we wish to determine the specific heat c_A of a body A of mass m_A. The body at a temperature T_A is immersed in a calorimeter which is full of water. The calorimeter and water are effectively body B and have the temperature T_B initially. When thermal equilibrium is reached, the final temperature is T_f. As the quantity of heat supplied by A is equal to that absorbed by B, we have

$$m_A c_A (T_A - T_f) = C_B (T_f - T_B)$$

From this equation the specific heat c_A may be deduced if the thermal capacity C_B of the calorimeter full of water is known.

This thermal capacity C_B can be determined in a preliminary experiment, for example by the method of balancing heat and electrical work described above. It can also be calculated from the specific heats c_i and the masses m_i of the various constituents of B, such as the water, parts of the calorimeter, the thermometer and so on, using

$$C_B = \sum m_i c_i$$

2.8 Application of the first law to perfect gases

We will use the first law of thermodynamics to develop our understanding of perfect gases.

INTERNAL ENERGY OF A PERFECT GAS

In general the internal energy of a fluid, being a state function, depends on two independent variables, for example temperature and volume. In the particular case of a perfect gas, there is no intermolecular potential energy that would depend on the volume available. As a result, whatever the structure of the molecules, the internal energy of a perfect gas depends only on the temperature. Under ordinary conditions U is directly proportional to T.[6]

For a monatomic gas the internal structure makes no contribution to the internal energy. Moreover, by definition, a perfect gas is in a rarefied state so that the interactions between the molecules are negligible. The internal energy U thus reduces to the kinetic energy of translation, U_t; according to our previous definition of temperature we have

$$U = \tfrac{3}{2}NkT$$

For a diatomic gas over a wide range of temperature, we find

$$U = \tfrac{5}{2}NkT$$

In addition to its mean translational energy $\tfrac{3}{2}kT$, each molecule has a mean rotational energy equal to kT. The theoretical explanation of this value of the rotational energy will be given in section 5.7.

JOULE'S EXPERIMENT

An experiment originally designed by Joule verifies that the internal energy U is effectively independent of the volume of the gas. Two connected vessels A and B are immersed in water in a calorimeter (Fig. 2.4). A gas contained

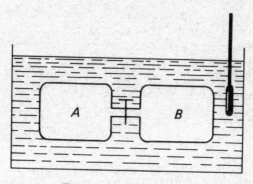

Figure 2.4. Joule's experiment.

in vessel A is allowed to expand into vessel B, which initially is evacuated. When equilibrium is established it is found that the water temperature is unchanged.

Although the volume of the gas has increased, the walls of the vessel have not moved so the gas has not exchanged work with its surroundings. The temperature of the water has not varied and therefore the gas has not

[6] This is a very special property of perfect gases. The relation between the internal energy and the temperature of any substance is generally more complicated.

exchanged heat with the surroundings. This means that the internal energy of the gas has not varied and confirms that at constant temperature the internal energy is independent of the volume.

In fact a small variation of temperature is observed because in practice the gas used in the experiment is not perfect.

Specific heats of a perfect gas

Let us consider unit mass of any substance. We have seen in section 2.6 that the specific heats at constant volume and constant pressure are respectively

$$c_V = \left(\frac{\partial U}{\partial T}\right)_V ; \qquad c_p = \left(\frac{\partial U}{\partial T}\right)_p + p\left(\frac{\partial V}{\partial T}\right)_p$$

Limiting ourselves to the case of a perfect gas we will calculate the difference $c_p - c_V$. As U depends only on T, we have

$$\left(\frac{\partial U}{\partial T}\right)_V = \left(\frac{\partial U}{\partial T}\right)_p = \frac{dU}{dT}$$

Moreover the equation of state for unit mass of a perfect gas is

$$pV = RT/M$$

where M is the mass of a mole; thus

$$\left(\frac{\partial V}{\partial T}\right)_p = \frac{R}{Mp}$$

Using these equations, the relation

$$M(c_p - c_V) = R = 8 \cdot 32 \, \text{JK}^{-1}$$

which was derived originally by Robert Mayer can be obtained.

In addition c_V can itself be calculated from the definition $c_V = (\partial U/\partial T)_V$ and the explicit form of U given above. For unit mass of a perfect gas, the internal energy has the form

$$U = \frac{l}{2}\frac{R}{M}T$$

where $l = 3$ for a monatomic gas, $l = 5$ for a diatomic gas (the relation is only approximate for a diatomic gas). Thus we have

$$Mc_V = \frac{l}{2}R$$

By using Mayer's relation the ratio of the specific heats is found to be a constant

$$\gamma = \frac{c_p}{c_V} = \frac{2 + l}{l}$$

REVERSIBLE ADIABATIC TRANSFORMATION OF A PERFECT GAS

During a reversible adiabatic transformation of a perfect gas, there is a relation between the pressure and the volume. To establish this relation, we first recall that the term 'adiabatic' signifies that the gas does not exchange heat with the surroundings; it is enclosed in an insulated container. The term 'reversible' implies that the volume of the gas varies slowly. For any fluid, by setting the heat absorbed during a reversible infinitesimal transformation to zero, we obtain[7]

$$0 = \delta Q = dU - \delta W = dU + p\,dV$$

$$= \left(\frac{\partial U}{\partial p}\right)_V dp + \left[\left(\frac{\partial U}{\partial V}\right)_p + p\right] dV$$

Thus p and V are related by a differential equation whose coefficients are known if the function $U(p, V)$ is known.

In the case of a perfect gas we have

$$U = \frac{l}{2}NkT = \frac{l}{2}pV; \quad \left(\frac{\partial U}{\partial p}\right)_V = \frac{l}{2}V; \quad \left(\frac{\partial U}{\partial V}\right)_p = \frac{l}{2}p$$

The differential equation involving p and V is simply

$$0 = \frac{l}{2}V\,dp + \left(\frac{l}{2} + 1\right)p\,dV$$

or alternatively by introducing the ratio γ of the specific heats ($\gamma = (l + 2)l$)

$$0 = V\,dp + \gamma p\,dV$$

Integrating this equation, the required relation, sometimes called Laplace's law, is obtained

$$pV^\gamma = C$$

where C is a constant.

By substituting for p in this equation using the equation of state $pV = NkT$, the relation expressing the variation of temperature as a function of the

---

[7] It should be recalled that it is necessary for the transformation to be reversible in order that the work done δW can be expressed in terms of the pressure of the fluid by the formula $p\,dV$.

57

volume during a reversible adiabatic transformation is found to be

$$T = \frac{C'}{V^{\gamma - 1}}$$

where C' is a constant. It can be seen that T is a decreasing function of V. Thus if a gas is compressed in a pump it tends to heat up. The microscopic mechanism is, of course, that the piston of the pump transfers kinetic energy to the gas molecules. Conversely, if the gas is allowed to expand and push against the piston, the gas cools.[8] This is the principle of Claude's method of liquefying air. Another application is the Wilson cloud chamber where the cooling produced by the expansion is sufficient to put the gas in a meta-stable state.

With an eye to the applications of Laplace's law it is useful to notice that it can also be written

$$p = A\rho^{\gamma}$$

where ρ is the density and A a constant.

Laplace's law plays a role in the calculation of the speed of sound in a gas. This question is treated in Appendix I.

Problems

2.1 A hammer of mass 0.2 kg falls 0.5 m on to an iron nail of mass 10^{-3} kg. Assuming that the nail absorbs all the kinetic energy of the hammer by friction, what is its increase in temperature? The atomic weight of iron is 56. Reading section 5.6 is advised.

2.2 The molar specific heat of magnesium at constant pressure can be expressed by

$$c_p = 25.7 + 6.26 \times 10^{-3}\,T - 3.3 \times 10^{-5}\,T^2$$

where c_p is in joules per mole per degree and T is the absolute temperature. What is the true specific heat of magnesium at 273 K? What is its mean specific heat between 273 K and 373 K?

2.3 At low temperatures the specific heat of sodium chloride at constant volume varies as

$$c_V = k\left(\frac{T}{\theta}\right)^3$$

[8] This expansion during which the gas pushes the piston back can be considered, approximately at least, as reversible. It must not be confused with the irreversible Joule expansion discussed earlier, which is not accompanied by any change in the temperature of a perfect gas.

where $k = 19.4 \times 10^2$ joules per mole per degree and θ is the Debye temperature of sodium chloride which has the value of 281 K.

(a) What is the true specific heat at constant volume at 20 K and at 40 K?
(b) What is the amount of heat required to raise the temperature of 1 mole from 20 K to 40 K?
(c) What is the mean specific heat between these two temperatures?

2.4 A real gas obeys Van der Waals' equation, which for unit mass of gas is

$$\left(p + \frac{a}{V^2}\right)(V - b) = rT$$

If its specific heat at constant volume is c_V and is independent of the temperature and pressure, express the specific heat at constant pressure in terms of the constants a, b, r, and c_V.

2.5 What is the internal energy $U(V, T)$ of the Van der Waals gas defined in problem 2.4?

2.6 What is the relation between the pressure p and the volume V of the Van der Waals gas in problem 2.4 during a reversible adiabatic expansion? (The relation $pV^\gamma = $ constant is valid only for a perfect gas.)

2.7 At the beginning of each of the following experiments a calorimeter contains 1 kg of water at 20°C.

(a) When 1 kg of water at 60°C is poured in, the final temperature is 38·30°C. What is the thermal capacity of the calorimeter?
(b) When 0·03 kg of ice at 0°C is added, the final temperature is 17·53°C. What is heat of fusion of ice?
(c) When 0·03 kg of ice at -10°C is placed in the calorimeter the final temperature is 17·41°C. What is the specific heat of ice between 0°C and -10°C?

2.8 A mixture of gas with the composition $2CO + O_2 + 2CO_2$ having a volume 7×10^{-4} m³ at 17°C and at atmospheric pressure is introduced into a bomb calorimeter (such a calorimeter is used to measure the heat of reaction of gases). The thermal capacity of the empty bomb is the same as 1 kg of water. The gas mixture is ignited and immediately after the explosion the internal pressure rises to 6·6 atm; after heat has been exchanged the temperature of the calorimeter reaches 17·58°C. Hence deduce

(a) the gas temperature just after the explosion,
(b) the mean molar specific heat of CO_2 at constant volume,
(c) the heat of reaction at constant volume for the combustion of 1 mole of CO.

2.9 An electric filament is placed in a calorimeter containing water. The whole apparatus has a thermal capacity of $1\ 000\ \text{J K}^{-1}$ and an initial temperature of 20°C. A d.c. voltage of 110 V is applied to the ends of the filament and after 10 min the temperature of the calorimeter is 25·32°C. Assuming that the resistance R of the filament varies according to the relation

$$R = R_0(1 + 0\cdot0043\ T)$$

where T is in degrees Celsius, deduce R_0, the filament resistance at 0°C.

2.10 A vehicle is propelled by compressed air from a cylinder initially at 200 atm and 20°C expanding into the atmosphere. If the vehicle is to have the same power (60 hp) and range (4 hours) as a car with an internal combustion engine, what mass of air is required? Assume the expansion is isothermal (1 hp = 745·7 W).

2.11 In a diesel engine the explosive mixture is sucked in at normal pressure and at a temperature of 15°C and compressed rapidly to a fifteenth of its original volume. What is the temperature and pressure when the compression is complete? Take $\gamma = 1\cdot3$.

2.12 In an air rifle a volume of $5 \times 10^{-6}\ \text{m}^3$ of air at 10 atm and 15°C expands adiabatically in the barrel of length 1 m and cross-sectional area $0\cdot25 \times 10^{-4}\ \text{m}^2$ and propels a bullet of mass $10^{-3}\ \text{kg}$ into the atmosphere.

What is the speed of the bullet as it leaves the barrel and the temperature and pressure of the air in the barrel at that instant?

2.13 Under certain conditions air at different altitudes in the atmosphere can be in reversible adiabatic equilibrium. Show that the temperature is then a linear function of the altitude. Experimentally a temperature gradient of 1 degree per 100 m is observed. Compare this with the theoretical value.

2.14 The experiment of Clément and Desormes. A reservoir closed by a tap contains a volume of air at room temperature T_0 and under a small excess pressure Δp_0 relative to the external pressure. The tap is opened for a short time so that the air expands adiabatically and reversibly to atmospheric pressure. The tap is then shut and, when thermal equilibrium has been established again, the excess pressure is found to be Δp_1. Show that the ratio $\gamma = c_p/c_V$ for air can be deduced from this experiment.

2.15 An evacuated container with adiabatic walls has a volume V. A hole is drilled in it so that it fills with air until the pressure inside equals the pressure outside. If the air was initially at a pressure p_0 and temperature T_0, what is the final temperature T_f? How much work has been supplied by the atmosphere during the transformation? If thermal contact between the inside of the container and the surroundings is now established, how much heat will flow through the container walls? For further practice calculate the answers taking $V = 10^{-3}$ m^3, $T_0 = 330$ K, and $p_0 = 1$ atm.

2.16 A vacuum pump removes 0·06 m^3 of gas per second measured at the pumping pressure from a reservoir of volume 0·2 m^3. How long will it take the pressure to go from $p_0 = 1$ atm to $p_1 = 10^{-3}$ mm Hg if the transformation is (a) isothermal and (b) adiabatic?

Calculate the mechanical energy supplied by the pump in both cases. What has become of this energy?

2.17 The enclosure shown in the figure has adiabatic outer walls. Pistons 1 and 2 can move freely without friction and divide the enclosure into three sections A, B, and C, which are filled with three equal masses of a perfect gas,

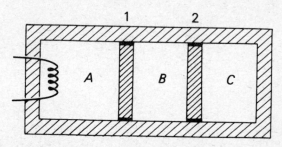

initially at the same temperature T_0 and pressure p_0. A resistance heats the gas in section A to a temperature T_A. What are the pressures p_A, p_B, and p_C and the temperatures T_B and T_C in the sections A, B, and C if

(a) both pistons are adiabatic partitions?
(b) piston 1 is adiabatic and piston 2 is diathermic?

Take $T_0 = 300$ K, $p_0 = 1$ atm, $T_A = 400$ K, $\gamma = 5/3$.

2.18 The enclosure shown in the figure has adiabatic outer walls. Pistons 1 and 2 are also adiabatic and can move freely without friction. A partition divides the enclosure into two sections A and B. Initially A is full of a perfect gas at P_0, V_0, and T_0 while B is evacuated. If a hole is drilled through the

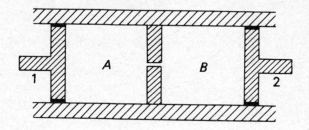

partition calculate the final temperatures and pressures in the two sections A and B in the following cases:

(a) With a diathermic partition, piston 1 is fixed and piston 2 is pushed back so that V_B goes from zero to V_0.

(b) Again with a diathermic partition, piston 1 moves so that $p_A = p_0$ always and piston 2 is fixed so that $V_B = V_0$.

(c) With adiabatic partition, piston 1 moves so that $p_A = p_0$ always and piston 2 is fixed so that $V_B = V_0$.

3. Entropy and the second law of thermodynamics

3.1 The second law of thermodynamics

In general, the natural development of macroscopic physical systems takes place in a preferred direction; a change in the opposite direction does not happen spontaneously. We will quote several examples to illustrate this.

(*a*) If two bodies with different temperatures are placed in contact with each other their temperatures become equal; on the other hand, if two bodies with the same temperature are in contact, there is never any spontaneous change in the temperature of one relative to the other.

(*b*) When a rotating wheel is stopped by application of the brakes, the brakes and the wheel rim become hot; while the brakes and the rim cool down, no spontaneous motion of the wheel in the opposite direction is ever observed.

(*c*) In Joule's experiment the gas flows out of the full vessel into the evacuated one; it never recompresses itself into the first vessel.

A natural process obeys the first law of thermodynamics, i.e., energy is conserved during the process. However, the inverse process also conserves energy. The first law says nothing about the direction in which the system develops.

In classical thermodynamics no attempt is made to explain this preferred direction starting from microscopic ideas. Instead a second law of thermodynamics is proposed *a priori* to account for the irreversibility of natural macroscopic processes.

There are several equivalent statements of the second law. Each of them asserts that some particular class of process is impossible. Starting from any one statement it can be shown that the other statements are in fact correct. As an example, we quote Kelvin's formulation of the law.

No engine can produce work from a single source of heat.

To illustrate this statement let us consider the operation of a steam engine. Such an engine takes heat from the boiler and transforms part of it into work, the rest of the heat being deposited into the condenser or the surrounding air. The engine works cyclically; once it has completed a cycle it is in the same state as it was at first. However, the operation of the engine has not *only* transformed heat taken from the boiler into work; it has also heated the condenser or the atmosphere.

An example of what the second law asserts is the impossibility of constructing a ship's engine which could produce work simply by cooling the sea water around it. Such an engine would produce what is called *perpetual motion of the second kind*;[1] the existence of such motion is thus forbidden by the second law.

The complete chain of logically connected results which make up thermodynamics can be deduced from only three laws (counting the zeroth law) without recourse to any microscopic considerations whatsoever. While this approach requires rather subtle reasoning, it has the merit among others of being particularly elegant. Although it follows the same path as the historical development of thermodynamics, it is a point of view that makes no attempt to explain the laws themselves.

In this book on the other hand we will continue to use the microscopic point of view of statistical mechanics. We will not put forward the second law *a priori*, but will attempt to show that it is a consequence of the behaviour of the microscopic constituents of matter.

3.2 Entropy and irreversibility. A simple example

ENTROPY AND DEVELOPMENT OF SYSTEMS IN TIME

Until now we have used such quantities as volume, pressure, temperature and internal energy for the macroscopic description of a system. Now we are going to introduce a new macroscopic quantity, *the entropy*, which will play an extremely important part in the rest of this book.

Let us consider a system whose macroscopic state is known. This is not sufficient to determine the microscopic state of the system since the same macroscopic state can result from very different microscopic states.

Let us assume for the moment that these microscopic states are discrete, i.e., they can be counted. Let Ω be the number of microscopic states corresponding to a given macroscopic state. The quantity $S = k \log \Omega$ is called the entropy of the system; k is a constant which, as we will see later, is

---

[1] 'Perpetual motion of the first kind' would be produced by an engine which works without a supply of energy of any form whatsoever. The first law asserts that this is impossible.

conveniently taken to be equal to Boltzmann's constant. This definition of entropy was first put forward by Boltzmann.

Even if a system is apparently at rest on a macroscopic scale, it is continually changing on a microscopic scale and can occupy various microscopic states in the course of time. Let us consider an isolated system which exchanges no energy whatsoever with the outside world. Because of the conservation of energy the microscopic states which the system can occupy have necessarily the same energy as the initial state; such states will be termed *accessible states*. For a macroscopic system the number of accessible microscopic states is astronomical and there is no question of being able to follow the development of the system in detail. Fortunately statistical information will be sufficient; we will only need to know the probability that a system can be found in such and such a microscopic state.

In the absence of any reason why the system should prefer one microscopic state to another, the following fundamental hypothesis, which is the basis of statistical mechanics, is made:

All accessible microscopic states of an isolated system have equal probability.

The probability concept can be understood here to relate to the development of the system in time. If a collection of accessible microscopic states is divided into groups, during the development of the system over a sufficiently long period, the time the system spends in each group is proportional to the number of states in that group.[2]

An example

We will illustrate these ideas by studying a simple system consisting of a collection of N atoms each with a spin $\frac{1}{2}$ and a magnetic moment μ. Each atom is bound in a fixed position. The projection of the spin on a quantization axis can only have the values $+\frac{1}{2}$ or $-\frac{1}{2}$ and, accordingly, the projection of the magnetic moment is either $+\mu$ or $-\mu$. Certain paramagnetic solids, for example a crystal of copper sulphate, can be considered as systems of this type.

Let us calculate the entropy of this system. Its macroscopic state is defined by its total magnetic moment along the quantization axis. Let the number of atoms oriented in the direction of this axis be $(N + n)/2$; since the total number of atoms is N, there are $(N - n)/2$ atoms oriented in the opposite sense. The total magnetic moment is then $n\mu$ and n is therefore the parameter which defines the macroscopic state.

However this state can be obtained Ω ways from the microscopic point of view. The $(N + n)/2$ atoms with a moment of $+\mu$ can be chosen arbitrarily

[2] The problem of proving this law of probability rigorously, starting from the laws of mechanics, is called the ergodic problem. It is only partially solved at the present time.

from among the N atoms. The number of possible choices is thus the number of ways of choosing $(N + n)/2$ objects from a total of N, i.e.,

$$\Omega = \frac{N!}{\left(\dfrac{N + n}{2}\right)! \left(N - \dfrac{N + n}{2}\right)!} = \frac{N!}{\left(\dfrac{N + n}{2}\right)! \left(\dfrac{N - n}{2}\right)!}$$

The corresponding entropy is

$$S = k \log \Omega = k \log \left\{ \frac{N!}{\left(\dfrac{N + n}{2}\right)! \left(\dfrac{N - n}{2}\right)!} \right\}$$

It is easily seen that Ω and therefore S reaches the maximum when $n = 0$. This maximum can be calculated using *Stirling's formula*,[3] which is valid for large x:

$$\log (x!) \sim x \log x - x + (\tfrac{1}{2}) \log (2\pi x) + \dots$$

For $n = 0$, we find

$$\frac{S}{k} = \log \Omega = \log \left\{ \frac{N!}{\left[\left(\dfrac{N}{2} \right)! \right]^2} \right\} \sim N \log 2 - \tfrac{1}{2} \log \frac{\pi N}{2}$$

For the macroscopic system that we are considering the number of atoms, N, is enormous (of the order of 10^{23}); $\log N$ is thus negligible compared with N so that $\log \Omega$ is smaller than $N \log 2$ by only a very small amount.

Let us now count the total number of microscopic states without the restriction that the magnetic moment has a given value. This number is $\Omega' = 2^N$ and so we have $\log \Omega' = N \log 2$.

So far we have obtained the following results. The number Ω of microscopic states corresponding to a given total magnetic moment is a maximum when the total magnetic moment is zero; what is more, since the entropy $k \log \Omega$ calculated for zero total magnetic moment is practically equal to that obtained by counting all possible microscopic states without any restriction on the value of the total magnetic moment, this maximum is very sharp.

We are now in a position to understand how our system is going to develop in time. We suppose that the system which is initially magnetized, i.e., in a state in which n is non-zero, is left to itself *in the absence of an external magnetic field*. Since every microscopic state without exception has zero energy, the system is free to make transitions from one microscopic state to

---

[3] Stirling's formula is explained in Appendix IV.

another without violating the principle of the conservation of energy. These transitions are caused by the small interactions which always exist between the atoms.

According to the fundamental postulate quoted above all microscopic states are equally probable. The probability of a macroscopic state is therefore proportional to the number Ω of microscopic states which correspond to it. As Ω displays a sharp maximum for the macroscopic state whose total magnetic moment is zero, it is most likely that the development of the system in time will bring it into this state. Subsequently the system has every chance of remaining in this state since it is so much more probable than the others. This macroscopic state is therefore the state of equilibrium.

In brief, the system develops without preference among the various microscopic states, but since the great majority of microscopic states correspond to the macroscopic state of equilibrium, the system seems to prefer this equilibrium state. Its development is irreversible.

During its development the system proceeds towards states of larger Ω with the result that the entropy $k \log \Omega$ increases, reaching its maximum when equilibrium is established. The entropy may then be calculated either by counting the microscopic states corresponding to the macroscopic state of equilibrium or what is effectively the same by counting all the accessible states without any restriction whatsoever.

GENERALIZATION

The considerations above can easily be generalized. In the very simplified example we have just discussed the macroscopic state of equilibrium was unique. Generally the parameters determining this state can vary. For these parameters it is convenient to take the internal energy U along with the necessary external parameters. When the system is a fluid, these parameters would be U and the volume V.

To enumerate the microscopic states, at least in principle, we can base our argument on quantum mechanics. Any finite system is characterized by quantum states with discrete energies which are in fact the microscopic states of the system. For a complex system such as the one we are considering, the energy levels are extremely close together and in the energy interval $(U, U + dU)$, corresponding for example to the smallest measurable interval, there is an enormous number Ω of states. The great majority of these states correspond to the macroscopic state of equilibrium; the entropy at equilibrium is therefore $k \log \Omega$. It can be considered as a function of U and the external parameters, i.e., as a state function.[4]

[4] Our definition of entropy is based on the quantum mechanical concept of energy levels. In practice the entropy can often be calculated without explicitly invoking quantum mechanics, classical mechanics being a sufficient approximation in numerous calculations.

Strictly the determination appears to depend on the choice of the interval dU. In fact it can be shown that in the limit where systems with a large number of particles are considered, S varies negligibly with dU.

For the same values of U and the external parameters there are fewer microscopic states corresponding to macroscopic states away from equilibrium and the entropy is smaller. By changing, the system has available to it a larger collection of accessible microscopic states. As a result of this *the entropy of an isolated system*, for which U and the other parameters are constants, *increases and attains its maximum when equilibrium is established*.

Entropy has the important property of being additive. Let us consider two systems A and B which are each in a known macroscopic state. System A can occupy Ω_A possible microscopic states and its entropy is $S_A = k \log \Omega_A$, while for B the corresponding quantities are Ω_B and S_B. The complete system $A + B$ possesses $\Omega_A \Omega_B$ possible microscopic states and its entropy is $S = k \log \Omega_A \Omega_B$. It can be seen that

$$S = S_A + S_B$$

and entropy is therefore an extensive quantity. In fact the definition of entropy is chosen to be $k \log \Omega$ rather than Ω itself in order to obtain an additive entropy.

3.3 Temperature. Pressure. The thermodynamic identity

TEMPERATURE

We are now in a position to review the concept of temperature.

The entropy S of a system in equilibrium is a function of its energy U and the external parameters, such as the volume V in the case of a fluid.

We will define the thermodynamic temperature T by the relation

$$\frac{\partial S}{\partial U} = \frac{1}{T}$$

where we have used a partial derivative since S also depends on the external parameters which are kept constant in this relation. We will see in the next section that the thermodynamic temperature coincides with the absolute temperature defined by the perfect gas; this is why we use the same notation T. First of all we will show that we have defined a quantity which has suitable properties for use as a temperature.

These characteristic properties are that for two bodies at different temtures heat flows from the hot body to the cold one and that two bodies in equilibrium have the same temperature.

To show that T has in fact these properties let us consider two bodies A and B in contact with one another, the complete system A and B being isolated, i.e., at constant internal energy. The volumes are fixed and the respective entropies are S_A and S_B so that the total entropy is $S = S_A + S_B$. The bodies A and B can exchange energy in the form of heat with the result that their entropies change. We will assume here that these changes occur sufficiently slowly so that both bodies remain in a state of internal equilibrium and S_A and S_B are well defined functions of their respective energies U_A and U_B.

Let dS be the variation of the total entropy during a small time interval. We have

$$dS = dS_A + dS_B = \frac{\partial S_A}{\partial U_A} dU_A + \frac{\partial S_B}{\partial U_B} dU_B$$

Because of the conservation of energy the energy given up by one body is absorbed by the other:

$$dU_B = -dU_A$$

Thus, using the definition of temperature, we find

$$dS = \left(\frac{1}{T_A} - \frac{1}{T_B} \right) dU_A$$

where T_A and T_B are the temperatures of the bodies A and B respectively.

As the complete system moves towards equilibrium the total entropy increases, i.e., $dS > 0$. Assuming that $T_A > T_B$, we then have $dU_A < 0$, i.e., body A loses energy. Thus heat does in fact flow from the body with the higher temperature to that with the lower temperature.

When equilibrium is established, the entropy S has reached its maximum. Since we can consider entropy as a function of U_A, this means that $dS/dU_A = 0$. Thus $T_A = T_B$ so that two bodies in thermal equilibrium have the same temperature.

The relation $\partial S/\partial U = 1/T$ therefore defines a scale of temperature with suitable properties. This temperature has an extremely fundamental character in as much as it represents for any body whatsoever an intrinsic property that is independent of a particular thermometer.

The possibility of defining an intrinsic thermodynamic temperature provides us with a justification of the zeroth law, which need no longer be postulated *a priori* (two bodies A and B which are in thermal equilibrium with a body C are in thermal equilibrium with each other). The 'law' arises from the simple consequence that $T_A = T_C$ and $T_B = T_C$ implies $T_B = T_A$. However we must emphasize the fact that *this simple argument is only possible because we knew the definition of the temperature T in the first place.*

PRESSURE

The entropy of a fluid is a function $S(U, V)$ of the internal energy U and the volume V. This relation can be inverted and U considered as a function of S and V. We will show that the pressure of the fluid is given by

$$p = -\left(\frac{\partial U}{\partial V}\right)_S$$

To do this we will consider an infinitesimal reversible adiabatic transformation of the fluid. Since $\delta Q = 0$, we have

$$dU = \delta W = -p\, dV$$

The difficulty in proving that $p = -(\partial U/\partial V)_S$ from this lies in the justification of the subscript 'S'; it is essential to be able to show that the entropy stays constant during the transformation.

We shall first show a more general result: *in an irreversible adiabatic transformation, the entropy of a system increases.* This statement is a generalization of the one which says that the entropy of an isolated system which evolves towards equilibrium increases. If the system is no longer isolated, and can exchange work (but not heat) with its surroundings, we can consider the larger isolated system $\varphi + \varphi'$ consisting of our system φ plus the external machinery φ' with which the work is exchanged. (For instance, a fluid in a vertical cylinder closed at its top by a piston can be compressed by loading the piston with a weight; the external machinery is the loaded piston and the earth which attracts it.) Now, $\varphi + \varphi'$ can be considered as isolated, and therefore its entropy increases as it evolves. The machinery φ' undergoes no microscopic change, and therefore its entropy is unchanged (in our example, the macroscopic displacement of the weight does not change its internal structure). The conclusion is that the entropy of the system φ increases.

In the limiting case when the adiabatic transformation is a reversible one, the entropy of the system stays constant. This can be shown as follows. The rate dS/dt at which the entropy S changes with time t is a function of the rate of change dV/dt of the volume. Let us expand dS/dt as a power series of dV/dt, for small values of dV/dt. Since dS/dt vanishes with dV/dt and is never negative, the leading term is the quadratic one

$$\frac{dS}{dt} = A\left(\frac{dV}{dt}\right)^2$$

Therefore

$$\frac{dS}{dV} = A\frac{dV}{dt}$$

For a reversible transformation, which is infinitely slow, dV/dt approaches zero, and thus dS/dV also approaches zero: in an adiabatic reversible transformation, the entropy stays constant.

We have therefore justified the subscript S in the formula

$$p = -\left(\frac{\partial U}{\partial V}\right)_S$$

THE THERMODYNAMIC IDENTITY

Gathering together the results of the preceding discussion, we can write the partial derivatives of the function $U(S, V)$ for a fluid as

$$\left(\frac{\partial U}{\partial S}\right)_V = T; \qquad \left(\frac{\partial U}{\partial V}\right)_S = -p$$

The differential of U is therefore

$$dU = T\,dS - p\,dV$$

This extremely fundamental relation is sometimes called the *thermodynamic identity*.

For a reversible transformation, $-p\,dV$ is the work δW done on the system; bearing in mind the first law $dU = \delta Q + \delta W$, we see that $T\,dS$ is the heat absorbed δQ and so

$$dS = \frac{\delta Q}{T}$$

for a reversible transformation.

The thermodynamic identity itself is valid whether the transformation is reversible or not since it expresses the variation of the state function U between two well-determined states (S, V) and $(S + dS, V + dV)$.

Finally it should be remarked that the relation $dS = \delta Q/T$ for a reversible transformation is valid for a system subjected to external forces other than the pressure on a fluid, for example anisotropic forces on a solid, electric, and magnetic fields and so on. As with the volume, the slow variation of the external parameters leaves the entropy constant and the previous argument can be generalized. Because of the variation of the external parameters, work is done on the system at constant entropy and so $\delta W = (dU)_S$. In a general reversible transformation

$$\delta W + \delta Q = dU = (dU)_S + \frac{\partial U}{\partial S}\,dS = \delta W + T\,dS$$

Comparing the first and last expressions we see that $\delta Q = T\,dS$.

Integrating the relation $dS = \delta Q/T$, we obtain for the entropy difference between two states (1) and (2)

$$S_2 - S_1 = \int_1^2 \frac{\delta Q}{T}$$

where the integral is taken along any reversible path whatsoever leading from state (1) to state (2). If the details of the heat absorbed along the whole path are known, the difference $S_2 - S_1$ can be calculated. The entropy S_2 is then obtained to within an additive constant.

Historically entropy was first defined by Clausius using the relation $S_2 - S_1 = \int_1^2 \delta Q/T$ (after it had been shown that the integral did not depend on the reversible path chosen). It was not until much later that Boltzmann proposed the microscopic interpretation $S = k \log \Omega$.

3.4 Perfect gases

In this section we will calculate explicitly the entropy S of a perfect gas and show that the thermodynamic temperature defined by $T = (\partial U/\partial S)_V$ does in fact lead to the relation $PV = NkT$ and therefore coincides with the absolute temperature defined starting from the perfect gas.

To calculate the entropy we must evaluate the number Ω of accessible microscopic states of the gas for a given macroscopic state defined by the energy U and the volume V. Ω may be obtained by counting the possible quantum states. However, except for the internal structure of the molecules, a gas is satisfactorily described by classical mechanics; Ω can be found just as well by counting the microscopic states defined by the positions and velocities of all the molecules (and in the long run by the parameters describing the internal state of each molecule).[5]

Since there is no interaction between the molecules of a perfect gas (except during collisions), the energy U depends only on the velocities and internal states of the molecules and not on their positions. Thus on the one hand we can count the number of possible configurations of the positions taking account of the single condition that the volume is V and on the other hand the number of possible configurations for the velocities and internal states of the molecules bearing in mind the single condition that the corresponding total energy is U.

First we will try to count the number of possible configurations for the positions of the molecules. The difficulty encountered here is that each

---

[5] A description of the internal structure of the molecules is necessary only for polyatomic molecules.

molecule has an infinite number of possible positions corresponding to every point in the volume V. In practice this turns out to be a minor difficulty which can be avoided by a trick. We divide the volume V into very small equal cells and specify the position of a molecule with limited precision, indicating only in which cell it is to be found. If τ is the volume of a cell, the volume V contains V/τ cells and a molecule has V/τ possible positions. The first molecule may thus be positioned in V/τ ways, the second molecule in V/τ ways,... the Nth molecule in V/τ ways. For the whole gas there are therefore $(V/\tau)^N$ possible configurations as far as the positions of the molecules are concerned.

Now let us turn to the number of possible configurations for the velocities and internal states of the molecules. We can avoid calculating this number explicitly here. It is enough for us to know that it depends on U, the total energy available to be distributed among the molecules (and not the volume which only governs the configurations of positions). Thus we let $f(U)$ be the number of possible configurations of the speeds and internal states of the molecules.

The most general microscopic state is obtained by associating any one of the $(V/\tau)^N$ configurations of position with any one of the $f(U)$ configurations of speeds and internal molecular states. The number of microscopic states is therefore the product $(V/\tau)^N f(U)$; this has the form $\Omega = CV^N f(U)$ where C is a constant.[6]

The entropy of a perfect gas is then

$$S = k \log \Omega = k[N \log V + \log f(U) + \log C]$$

and its differential is

$$dS = k\left[\frac{N}{V}dV + \frac{d[\log f(U)]}{dU}dU\right]$$

Rearranging the thermodynamic identity we have

$$dS = \frac{p}{T}dV + \frac{1}{T}dU$$

where T is the thermodynamic temperature. Equating the coefficients of V in these last two equations gives immediately

$$T = \frac{pV}{Nk}$$

where T is again the thermodynamic temperature. On the other hand

---

[6] The value chosen for τ only occurs in the multiplicative constant C so we do not attempt to calculate it here.

however, (pV/Nk) is by definition the absolute temperature on the perfect gas scale. Thus we have just shown that the thermodynamic temperature and absolute temperature defined by the perfect gas are identical.

It should be noticed that the identification is complete only because we have taken the constant k in the definition of entropy $S = k \log \Omega$ as Boltzmann's constant R/N_0.

We saw in section 1.4 that the temperature on the perfect gas scale is directly related to the mean kinetic energy of translation of a molecule

$$\tfrac{1}{2}m\overline{v^2} = \tfrac{3}{2}kT$$

The identification of T with the thermodynamic temperature provides us with a valid demonstration of the equipartition of energy between two different perfect gases in thermal equilibrium. On account of the general properties of the thermodynamic temperature, discussed in section 3.3, the two gases have the same thermodynamic temperature and in consequence the same kinetic energy $\tfrac{1}{2}m\overline{v^2}$.

Comparison of the coefficients of dU in the two expressions for dS given above supplies further interesting information. We find that

$$k\frac{d[\log f(U)]}{dU} = \frac{1}{T}$$

Thus the temperature is a function of the internal energy or inversely, the internal energy of a perfect gas depends only on its temperature. The quite general methods of the present chapter have enabled us to reproduce this very remarkable property of perfect gases. If we had calculated the function $f(U)$ explicitly we would have obtained the explicit relation between U and T.

For our purposes, however, we will make use of the relation we know already

$$U = \tfrac{3}{2}NkT$$

to determine the function $f(U)$. This relation is valid only for a perfect *monatomic* gas so we will limit ourselves to that case for the rest of this section. We find that

$$\frac{d[\log f(U)]}{dU} = \frac{1}{kT} = \frac{3N}{2U}$$

and by integration this becomes

$$\log f(U) = \frac{3N}{2} \log U + C'$$

where C' is a constant. If we substitute this result into the expression for the entropy we obtain

$$S = k\left[N \log V + \frac{3N}{2} \log U + \log C \right]$$

where the constant C' has been included in the constant C. To within an additive constant this represents an explicit expression for the entropy of a perfect monatomic gas.

Since $S = k \log \Omega$, we can recover the expression for the number of accessible microscopic states:

$$\Omega = C V^N U^{3N/2}$$

It is impossible, however, to obtain the value of C except by counting the accessible quantum states explicitly, which we will not do here.

3.5 General qualitative characteristics of entropy and temperature

We have just seen that as a function of energy the number of accessible microscopic states behaves like $U^{3N/2}$ for a perfect monatomic gas.

It turns out that there is an analogous behaviour for almost all systems occurring in nature. Ω is *very approximately* proportional to $(U - U_0)^{\alpha N}$, where U_0 is the lowest possible energy of the system and α is a number of the order of unity. For a macroscopic system the number N of molecules is of the order of 10^{23}. As a result Ω increases extremely rapidly as a function of U. The basic reason for this behaviour of the function $\Omega(U)$ is that as soon as U increases the number of ways in which the total energy of excitation $(U - U_0)$ can be constructed with the different microscopic constituents of the system becomes considerable if the number N of constituents is large.

The statement that Ω behaves as $(U - U_0)^{\alpha N}$ must not be taken too seriously when U is very close to U_0. For $U = U_0$, Ω is not zero but small. At the minimum energy U_0 there is very often only a single quantum state called the *ground state*, and so $\Omega(U_0) = 1$. It can happen that the ground state is *degenerate*, i.e., there are several quantum states associated with U_0, but the number $\Omega(U_0)$ of these states is always small.

The behaviour of entropy and temperature as a function of energy can be inferred from the behaviour of Ω. We have

$$S = k \log \Omega \sim \alpha N k \log (U - U_0) + \text{constant}$$

and

$$\frac{1}{T} = \frac{\partial S}{\partial U} \sim \frac{\alpha N k}{U - U_0}$$

or rearranging

$$kT \sim \frac{U - U_0}{\alpha N}$$

The entropy and the temperature are both increasing functions of energy. The temperature T is a positive quantity.[7] The energy of excitation per particle is of the order of kT.

When the energy tends towards its minimum value U_0, the temperature T tends towards zero; this limiting temperature is called absolute zero. The corresponding limiting value for the entropy is $k \log \Omega(U_0)$; as $\Omega(U_0)$ is unity or a small number, the limiting value of the entropy is zero or at the most is of the order of $k \sim 10^{-23}$ joules per degree, which is negligible from the macroscopic point of view.

This last result is the content of Nernst's law: **when the temperature of a system tends to absolute zero, its entropy tends to zero.** Nernst's law is often called the *third law of thermodynamics*. From the microscopic point of view it arises as a simple consequence of the absence of other energy levels in the region of the ground state.

The expression for the entropy of a perfect gas given in section 3.4 seems to be in conflict with Nernst's law, since when $T = 0$ we have $U = 0$, which makes $S = -\infty$. In reality the perfect gas laws cannot be extrapolated to low temperatures. On the one hand all known bodies, with the exception of helium, solidify and the concept of a gas is no longer relevant. On the other hand we have established the perfect gas laws without taking account of the quantum effects which become important at low temperatures.[8]

As an example of the application of Nernst's law we will consider the change of entropy of a body between absolute zero and the temperature T. Since the entropy is a state function, the change does not depend on the way in which the transformation occurs. If the transformation is assumed reversible, the relation $dS = \delta Q/T$ may be used. We find by integration

$$S(T) - S(0) = \int_0^T \frac{\delta Q}{T} = \int_0^T \frac{C(T) \, dT}{T}$$

where $C(T)$ is the thermal capacity under the conditions which apply to the transformation. According to Nernst's law, $S(0) = 0$. On the other hand $S(T)$ has a finite value. Thus the integral over T must be convergent, which is

[7] See however section 5.3.

[8] The wave nature of molecules and quantum effects can no longer be ignored when the de Broglie wavelength associated with a molecule is comparable with or greater than the mean distance between molecules. At a low temperature the molecules have a small energy and momentum p and thus the de Broglie wavelength $\lambda = h/p$ is not negligible.

only possible in the region of $T = 0$ if $C(T)$ tends to zero with T. From this we conclude that a thermal capacity necessarily tends to zero at low temperatures. Once again the laws relating to perfect gases, which predict constant thermal capacities, cannot be extrapolated to low temperatures.

3.6 The change of entropy in irreversible processes

In the case of an infinitesimal reversible transformation we have established the relation $dS = \delta Q/T$. Now we will show that during an irreversible transformation a system which absorbs heat Q from a source at temperature T undergoes a change of entropy ΔS such that $\Delta S > Q/T$.

We should remark immediately that the system is not in equilibrium during an irreversible process and in general does not have a definite temperature; the T which appears in the inequality $\Delta S > Q/T$ is the temperature of the *source* of heat, which is assumed to be in a state of equilibrium internally. This distinction is unnecessary in the case of a reversible transformation for then the system and source have infinitely close temperatures.

First let us consider a system which undergoes an irreversible adiabatic transformation so that $Q = 0$. We have already seen in section 3.3 that the entropy increases.

An extreme example of an irreversible adiabatic transformation is the Joule expansion of a gas from vessel A to vessel B without exchange of heat with the surroundings. If we recall that for a perfect gas the internal energy does not change, examination of the expression for the entropy of a perfect gas given in section 3.4 reveals that the increase in entropy is caused by the increase in the term $Nk \log V$. It can be understood intuitively that the entropy has increased because of the greater number of microscopic configurations of positions available in the larger volume.

We will now turn to the case of any irreversible transformation with the condition however that any exchanges of heat are made only with a source of well-defined temperature T. In addition the system can exchange work with the surroundings. The system and the source together do not exchange heat with the rest of the universe and therefore undergo an adiabatic (irreversible) transformation so that their entropy increases. Let ΔS be the change in the entropy of the system and $\Delta S'$ that of the source. The total change is positive so that

$$\Delta S + \Delta S' > 0$$

If the system absorbs heat Q from the source, the source absorbs $-Q$. A source is a sufficiently large system not to be perturbed by the heat it absorbs or loses. Thus the source remains in equilibrium at the same temperature T and undergoes a transformation which can be considered reversible giving

$\Delta S' = -Q/T$. The required inequality for an irreversible transformation

$$\Delta S > Q/T$$

follows from this.

This reasoning can be generalized to the case of a system which is put successively in contact with several sources. Once again, the entropy of the system and all the sources combined increases and the variation of the entropy of the system for an irreversible transformation is

$$\Delta S > \sum_i \frac{Q_i}{T_i}$$

where T_i denotes the temperature of the ith source and Q_i the heat absorbed from it by the system. In the limit in which the system is placed in contact with an infinite number of sources of various temperatures we have

$$\Delta S > \int \frac{\delta Q}{T}$$

where δQ denotes the infinitesimal amount of heat absorbed from the source with temperature T.

The variation of entropy of a system in both reversible and irreversible transformations can be condensed into the single expression

$$\Delta S \geqslant \int \frac{\delta Q}{T}$$

where the equality only holds if the transformation is reversible.

It can be seen that as soon as irreversible phenomena take place, they are accompanied by an increase in entropy which must be added to the change $\int \delta Q/T$ and leads to the inequality. In a rather simple-minded way we can say that the entropy of a system, which is a measure of the disorder of its state, can be increased either by the transfer of heat, which is a disordered form of energy, or by disturbing it violently by making it undergo a transformation in a finite time.

The inequality $\Delta S \geqslant \int \delta Q/T$ represents a mathematical expression of the second law of thermodynamics. If the transformation is *cyclic*, i.e., the system is returned to its initial state, $\Delta S = 0$ and we obtain the inequality due to Clausius

$$\oint \frac{\delta Q}{T} \leqslant 0$$

In particular if the system has been in contact with a single source of temperature T, the Clausius inequality reduces to $Q/T \leqslant 0$. Thus the system can only lose heat. According to the first law it can only have received work since it has returned to its initial state. The system cannot be of use as an engine. Thus we have arrived at the statement of the second law proposed *a priori* by Kelvin.

No engine can produce work from a single source of heat.

3.7 The efficiency of thermal machines

ENGINES

We have just seen that a cyclic machine cannot produce work from a single source of heat. If two sources of heat with different temperatures T_1 and T_2 are available, work may be produced with the help of a cyclic machine, but there is a natural maximum for the efficiency that can be expected.

Assuming that $T_1 > T_2$, we will consider a machine which in the course of one cycle absorbs heat Q_1 by contact with the hot source T_1 and gives up heat Q_2 by contact with the cold source T_2; the difference $W = Q_1 - Q_2$ is converted into the work the machine produces.[9] The ratio $\eta = W/Q_1$ is called the *efficiency*; from the engineer's viewpoint the heat Q_2 is lost and the problem is to produce as much work W as possible from the smallest possible quantity of heat Q_1.

For this example the Clausius inequality can be written

$$\oint \frac{\delta Q}{T} = \frac{Q_1}{T_1} - \frac{Q_2}{T_2} \leqslant 0$$

Thus

$$\eta = \frac{Q_1 - Q_2}{Q_1} = 1 - \frac{Q_2}{Q_1} \leqslant 1 - \frac{T_2}{T_1}$$

The inequality

$$\eta \leqslant 1 - \frac{T_2}{T_1}$$

is known as *Carnot's theorem*: whatever the technical details of the construction of the engine, its efficiency cannot exceed the value $1 - (T_2/T_1)$. This limit is attained in the ideal case in which the engine operates reversibly (functions infinitely slowly, no friction, and so on). The efficiency of a real machine is always less than the Carnot limit.

[9] For this section, and this section only, we have adopted the particular sign convention that Q_1 is the heat absorbed, Q_2 the heat given up and so on.

REFRIGERATORS

A refrigerator is a machine which absorbs heat Q_2 in contact with a cold source at temperature T_2 as well as work W; it gives up a total $Q_1 - Q_2 + W$ in contact with the hot source of temperature T_1 (in a domestic refrigerator the ice box is the cold source, the surroundings of the refrigerator is the hot source, and the work is supplied by electricity from the mains). The performance is measured by the coefficient $\eta' = Q_2/W$.

Here the Clausius inequality can be written

$$\oint \frac{\delta Q}{T} = \frac{Q_2}{T_2} - \frac{Q_1}{T_1} \leqslant 0$$

Thus

$$\frac{1}{\eta'} = \frac{Q_1 - Q_2}{Q_2} = \frac{Q_1}{Q_2} - 1 \geqslant \frac{T_1}{T_2} - 1 = \frac{T_1 - T_2}{T_2}$$

or

$$\eta' \leqslant \frac{T_2}{T_1 - T_2}$$

Since the coefficient of performance $\eta' = Q_2/W$ has an upper limit, it is impossible for a refrigerator to function without a supply of external work W. This result is the same as the statement of the second law proposed *a priori* by Clausius.

A process, whose only effect would be to transfer heat from a cold source to a hot source, is impossible.

Problems

3.1 If two masses of water M_1 and M_2 with temperatures T_1 and T_2 are mixed, find the final temperature T and the change in entropy during the mixing.

3.2 Two monatomic perfect gases each occupying a volume V, have different temperatures T_1, T_2, and different pressures p_1, p_2. If they are allowed to mix, what are the final temperatures and pressures? What is the change in entropy on mixing.

3.3 Repeat problem 3.2 using equal masses of perfect gases at different temperatures and pressures instead of equal volumes.

3.4 A domestic refrigerator consumes 50 W of electric power. The temperature of the evaporator is held at $-20°C$ and that of the radiator at $40°C$.

Assuming that the coefficient of performance is 50 per cent of the maximum given by Carnot's relation, calculate the quantity of heat absorbed by the evaporator in an hour.

3.5 Heat pump. To keep the temperature inside a house at 20°C when the outside temperature is $-10°C$, 200 MJ must be supplied per hour. What would be the electric power consumed by a refrigerator with an efficiency equal to the theoretical maximum, cooling the outside and heating the inside. Discuss the economics of the question (1 kg of fuel oil gives out 40 MJ on burning and costs about 1·5p. 1 kW of electricity cost 0·8p in 1971).

3.6 A 10 Ω resistance has a current of 10 A passing through it and is maintained at 30°C by a current of air. What is the variation of entropy per second of (a) the resistance, (b) the current of air?

3.7 A system consisting of a large number N of atoms of spin $\frac{1}{2}$ and magnetic moment μ is placed in a field of constant magnetic induction B. Assume that the total energy U of a system with spin in a field **B** is $U = -\mathbf{M} . \mathbf{B}$, where **M** denotes the total magnetic moment of the system, and that if N is large $\log N! = N \log N - N$.

(a) If Ω denotes the number of possible states with energy U, calculate $\log \Omega$ as a function of U.
(b) Use the definition of T, i.e., $(kT)^{-1} = \partial (\log \Omega)/\partial V$ to find the relation between the energy U and the temperature T of the system of spins.
(c) Deduce a relation between B and T for an isolated system of spins which are not undergoing any interaction.

3.8 If two equal masses of water with temperatures T_1 and T_2 ($T_1 < T_2$) are available, what is the maximum amount of work that could be produced by a thermal machine of maximum efficiency, operating with these two masses of water as hot and cold sources? What would be the final temperature of the system?

3.9 A thermal machine operates on a Carnot cycle (two isotherms and two adiabatic curves) and utilizes a gas which obeys the equation of state $p(V - b) = RT$ (the equation of Clausius). Show that its efficiency is the same as that of a machine using a perfect gas.

3.10 A thermal machine operates by using a mass of water, M, initially at temperature T_1, as a cold source and a hot source maintained at temperature T_0 by a thermostat. What is the maximum amount of work it can produce?

3.11 What is the minimum amount of work that must be supplied to a mole of a monatomic perfect gas at an initial pressure p_0 and temperature T_0 to raise its pressure to p_1 at the same temperature T_0?

3.12 A mole of a monatomic perfect gas at a temperature T_0 is compressed isothermally from $p_0 = 1$ atm to $p_1 = 50$ atm. Then it is allowed to expand isentropically until it reaches the pressure p_0. What is the temperature T? What is the change in entropy ΔS_1 of the gas during the transformation?

The same sequence of operations is carried out n times starting from the temperature obtained at the end of the preceding transformation. What is the final temperature T_n obtained and the change in entropy after n operations?

This result seems to contradict the third law of thermodynamics ($S = 0$ if $T = 0$). However we should ask ourselves whether the hypothesis of the perfect gas is valid at low temperatures.

3.13 Show that in the (p, V) plane, the curves representing two isentropic transformations of any fluid never intersect one another.

3.14 A mole of monatomic perfect gas is taken around a cycle consisting of two transformations at constant volume (isochoric processes) and two transformations at constant pressure (isobaric processes). The gas starts at (p_1, V_1) and passes through $(p_2, V_1), (p_2, V_2)$ and (p_1, V_2) before returning to (p_1, V_1). What is the work done on the gas and the heat absorbed? What is the efficiency relative to a Carnot engine operating between the extreme temperatures of the cycle?

4. Some developments and applications of classical thermodynamics

4.1 Aims of classical thermodynamics

Basing our arguments on our knowledge of the microscopic structure of matter, we have defined such concepts as heat, temperature, internal energy, and entropy and have found relations between them. Historically, these concepts were introduced from a purely macroscopic point of view. Certainly equations such as $dU = \delta W + \delta Q$ and $\delta Q = T\,dS$ for a reversible process, which summarize the fundamental laws of classical thermodynamics, have a very general validity, independent of the detailed microscopic structure.

The object of classical thermodynamics is to establish relations between various macroscopic properties of a particular system using these laws. Thermodynamics constitutes an extremely powerful tool for investigating a system, since it enables information to be obtained which would otherwise require a knowledge of the mechanism of the phenomena governing the system. However, it is impossible to calculate the properties of a system without any knowledge of its behaviour. Thermodynamics on its own can only supply *relations* between different phenomena.

The present chapter will be devoted to some examples of the application of the methods of classical thermodynamics.

4.2 The thermodynamic functions and Maxwell's relations

MAXWELL'S RELATIONS AND THE DEFINITION OF THE THERMODYNAMIC FUNCTIONS

The thermodynamic identity for fluids

$$dU = T\,dS - p\,dV$$

tells us that

$$\left(\frac{\partial U}{\partial S}\right)_V = T \quad \text{and} \quad \left(\frac{\partial U}{\partial V}\right)_S = -p$$

Differentiating these equations with respect to V and S respectively and using $\partial^2 U/\partial S \partial V = \partial^2 U/\partial V \partial S$ we find that

$$\left(\frac{\partial T}{\partial V}\right)_S = -\left(\frac{\partial p}{\partial S}\right)_V$$

This is one of Maxwell's relations. Let us see what we can learn from such a relation. Thermodynamics on its own does not enable us to calculate explicitly the way in which a system behaves, but it does relate the behaviour of one system to that of another. For example consider the process in which a substance is heated at constant volume so that its pressure increases and the process in which a substance is compressed in an adiabatic manner so that its temperature rises. Since $T\,dS = \delta Q$, Maxwell's relation given above may be written

$$\left(\frac{dT}{dV}\right)_{\text{adiabatic}} = -T\left(\frac{dp}{\delta Q}\right)_V$$

Thus if the temperature rise dT for a small volume change dV is known for the adiabatic process, the pressure increase dp for a small transfer of heat δQ can be calculated for the process at constant volume.

The thermodynamic identity and Maxwell's relation which follows from it use entropy and volume as independent variables. If the system is specified by another pair of variables, it is of interest to replace U by another *thermodynamic function*, suited to the choice of variables. These functions are constructed so that their differentials are easily expressed in terms of small increases in the chosen variables.

If the independent variables are temperature and volume, the function to consider is the *free energy*:[1]

$$F = U - TS$$

Thus we have

$$dF = dU - T\,dS - S\,dT$$

and bearing in mind the thermodynamic identity this becomes

$$dF = -S\,dT - p\,dV$$

so that dF is expressed in terms of dT and dV.

---

[1] Sometimes this is called the *Helmholtz free energy* to avoid confusion with another function, the *Gibbs free energy*.

At constant temperature we simply find $-dF = p\,dV$; under such conditions the decrease in F is equal to the work done in a reversible transformation and this is why F is called the free energy.

From the form of dF, we can immediately conclude that

$$\left(\frac{\partial F}{\partial T}\right)_V = -S \quad \text{and} \quad \left(\frac{\partial F}{\partial V}\right)_T = -p$$

Differentiating these equations with respect to V and T respectively we find the corresponding Maxwell relation

$$\left(\frac{\partial S}{\partial V}\right)_T = \left(\frac{\partial p}{\partial T}\right)_V$$

or

$$\frac{1}{T}\left(\frac{\delta Q}{dV}\right)_T = \left(\frac{\partial p}{\partial T}\right)_V$$

If the independent variables are entropy and pressure we construct the *enthalpy*

$$H = U + pV$$

which is such that

$$dH = T\,dS + V\,dp$$

The corresponding Maxwell relation is

$$\left(\frac{\partial T}{\partial p}\right)_S = \left(\frac{\partial V}{\partial S}\right)_P$$

or

$$\left(\frac{dT}{dp}\right)_{\text{adiabatic}} = T\left(\frac{dV}{\delta Q}\right)_P$$

Finally, if the independent variables are temperature and pressure, the *thermodynamic potential* or the *Gibbs free energy* is introduced

$$G = U + pV - TS$$

which yields

$$dG = -S\,dT + V\,dp$$

The corresponding Maxwell relation is

$$\left(\frac{\partial S}{\partial p}\right)_T = -\left(\frac{\partial V}{\partial T}\right)_P$$

or

$$\frac{1}{T}\left(\frac{\delta Q}{dp}\right)_T = -\left(\frac{\partial V}{\partial T}\right)_p$$

MACROSCOPIC CALCULATION OF THE THERMODYNAMIC FUNCTIONS

It is impossible to calculate the thermodynamic functions starting only from the equation of state of a substance. However, if in addition to the equation of state, a thermodynamic function for a restricted suitable set of states is known, then the thermodynamic functions for all possible states can be deduced.

For example, let us consider the free energy. By integrating the relation

$$\left(\frac{\partial F}{\partial V}\right)_T = -p$$

we obtain

$$F(V, T) = -\int_{V_0}^{V} p(V', T)\,dV' + F(V_0, T)$$

Thus if $F(V_0, T)$ is known at volume V_0 and all temperatures and also the equation of state, F can be calculated for all other states (V, T).

By differentiation, the entropy is obtained

$$S(V, T) = -\left(\frac{\partial F(V, T)}{\partial T}\right)_V = \int_{V_0}^{V} \frac{\partial p(V', T)}{\partial T}\,dV' + S(V_0, T)$$

The internal energy $U = F + TS$ is given by

$$U(V, T) = \int_{V_0}^{V}\left(-p + T\frac{\partial p}{\partial T}\right)dV' + U(V_0, T)$$

Classical thermodynamics alone is not sufficient to allow us to make further progress. The equation of state giving $p(V, T)$ and the function $U(V_0, T)$ for a volume V_0 must be known. Here we will complete the calculation of the internal energy by taking V_0 infinite; the gas is then infinitely rarefied and its internal energy is that of a perfect gas. Thus we have

$$U(V_0, T) = U(\infty, T) = \frac{l}{2}NkT$$

where $l = 3$ for a monatomic gas and $l = 5$ for a diatomic gas over a wide range of temperatures. To calculate the integral we will assume that Van der Waals' equation is the equation of state (this involves an approximation).

$$p = \frac{NkT}{V - B} - \frac{A}{V^2}$$

Then we obtain

$$-p + T\frac{\partial p}{\partial T} = \frac{A}{V^2}$$

and

$$U(V, T) = \int_\infty^V \frac{A}{V'^2}\, dV' + \frac{l}{2} NkT$$

which gives finally

$$U = \frac{l}{2} NkT - \frac{A}{V}$$

It can be seen that the internal energy contains the term $-A/V$ in addition to the energy $\frac{1}{2}l\, NkT$ of a perfect gas. This term $-A/V$ represents the potential energy due to the interaction of the molecules.

A REMARK ON THE MICROSCOPIC POINT OF VIEW

The example we have just treated was intended to illustrate a typical application of classical thermodynamics; the equation of state was assumed known, for example from experiment, and was used along with the function $U(\infty, T)$ to calculate the thermodynamic function $U(V, T)$.

Statistical physics adopts the inverse procedure. The required thermodynamic functions are determined in a microscopic way and the equation of state deduced from them. For example, Van der Waals' equation is obtained by the following procedure. Let $F(V, T) = U - TS$ be the free energy of a fluid. In the first approximation, the attractive forces between the molecules are neglected and the molecules considered as simple hard spheres. In this case we will write the free energy as $F^{(0)}(V, T)$. In the second approximation, the attractive forces are taken into account, but are considered so small that they do not disturb the configuration of the molecules. The entropy S is thus unchanged while the energy U acquires a term representing the potential energy due to the attraction between the molecules. Each molecule interacts with a number of neighbouring molecules. Since this number is approximately proportional to the density of the fluid, the potential energy due to the interaction is of the form $-A/V$, where A is a constant and V the volume of the fluid. Thus the total free energy is

$$F = F^{(0)} - \frac{A}{V}$$

Now we can find the pressure

$$p = -\left(\frac{\partial F}{\partial V}\right)_T = -\left(\frac{\partial F^{(0)}}{\partial V}\right)_T - \frac{A}{V^2}$$

87

$-(\partial F^{(0)}/\partial V)_T$ is the pressure of a gas consisting of hard spheres and this may be reasonably approximated by $NkT/(V - B)$. Substitution of this into the expression for p, produces Van der Waals' equation

$$p = \frac{NkT}{V - B} - \frac{A}{V^2}$$

4.3 Thermal coefficients of a fluid

The heat δQ absorbed by a fluid which undergoes a *reversible* infinitesimal transformation is a definite quantity. Since the work δW done on the fluid has the well-defined value $-p\,dV$, the heat absorbed is $\delta Q = dU - \delta W = dU + p\,dV$ and is thus expressed in terms of state variables and their variations, which have definite values.

To simplify the expressions, in the present section we will consider a unit mass of fluid. Depending on the choice of independent variables, we can write

$$\delta Q = c_V\,dT + l\,dV = c_p\,dT + h\,dp = \lambda\,dp + \mu\,dV$$

The coefficients $c_V, l, c_p, h, \lambda, \mu$ are state functions called *thermal coefficients*. (If $dV = 0$, $\delta Q = c_V\,dT$; if $dp = 0$, $\delta Q = c_p\,dT$. c_V and c_p should be recognized as the specific heats at constant volume and constant pressure respectively.)

We intend to establish the relations which the thermal coefficients obey, assuming the equation of state of the fluid is known.

RELATIONS INDEPENDENT OF THERMODYNAMICS

By a simple change of variables, we find[2]

$$\delta Q = c_V\,dT + l\,dV = c_V\,dT + l\left(\frac{\partial V}{\partial T}\,dT + \frac{\partial V}{\partial p}\,dp\right)$$

which, by comparison with $\delta Q = c_p\,dT + h\,dp$, gives the relations

$$c_p = c_V + l\frac{\partial V}{\partial T} \quad \text{and} \quad h = l\frac{\partial V}{\partial p}$$

---

[2] When the partial derivative of one of the variables p, V, or T is taken with respect to another, it goes without saying that the third remains constant. Thus sometimes an abbreviated notation, such as $\partial V/\partial T$ for $(\partial V/\partial T)_p$, is used.

These can be rearranged[3] yielding

$$l = (c_p - c_V)\frac{\partial T}{\partial V} \quad \text{and} \quad h = -(c_p - c_V)\frac{\partial T}{\partial p}$$

An analogous argument based this time on the comparison of

$$\delta Q = c_V \, dT + l \, dV = c_V\left(\frac{\partial T}{\partial p} \, dp + \frac{\partial T}{\partial V} \, dV\right) + (c_p - c_V)\frac{\partial T}{\partial V} \, dV$$

$$= c_V \frac{\partial T}{\partial p} \, dp + c_p \frac{\partial T}{\partial V} \, dV$$

with $\delta Q = \lambda \, dp + \mu \, dV$ gives

$$\boxed{\lambda = c_V \frac{\partial T}{\partial p}} \quad \text{and} \quad \boxed{\mu = c_p \frac{\partial T}{\partial V}}$$

Thus $l, h, \lambda,$ and μ have been expressed as functions of only the specific heats c_p and c_V and quantities calculable from the equation of state.

RELATIONS DERIVED FROM THERMODYNAMICS

Thermodynamics enables us to calculate $l, h, c_p - c_V$ as well as the derivatives $(\partial c_V/\partial V)_T$ and $(\partial c_p/\partial p)_T$.

Two of Maxwell's relations give l and h immediately. We have

$$\left(\frac{\partial p}{\partial T}\right)_V = \left(\frac{\partial S}{\partial V}\right)_T = \frac{1}{T}\left(\frac{\delta Q}{dV}\right)_T = \frac{l}{T}$$

where the last equality follows from the definition of l, and

$$\left(\frac{\partial V}{\partial T}\right)_p = -\left(\frac{\partial S}{\partial p}\right)_T = \frac{1}{T}\left(\frac{\delta Q}{dp}\right)_T = -\frac{h}{T}$$

the last equality being derived from the definition of h. We have now obtained Clapeyron's relations.[4]

$$\boxed{\left(\frac{\delta Q}{dV}\right)_T = l = T\frac{\partial p}{\partial T}} \quad \text{and} \quad \boxed{\left(\frac{\delta Q}{dp}\right)_T = h = -T\frac{\partial V}{\partial T}}$$

---

[3] Here we have used the following properties of the partial derivatives

$$\frac{\partial V}{\partial T} = 1 \bigg/ \left(\frac{\partial T}{\partial V}\right) \quad \text{and} \quad \left(\frac{\partial T}{\partial V}\right)\left(\frac{\partial V}{\partial p}\right) = -\frac{\partial T}{\partial p}$$

(notice the sign of the last identity).

[4] Translator's note: These relations are not usually given a special name in the English literature.

While discussing the relations independent of thermodynamics we saw that

$$c_p - c_V = l\frac{\partial V}{\partial T}$$

Substituting for l from Clapeyron's relations, we find

$$c_p - c_V = T\frac{\partial p}{\partial T}\frac{\partial V}{\partial T} = -T\frac{\partial p}{\partial V}\left(\frac{\partial V}{\partial T}\right)^2$$

This result, which is valid for any fluid whatsoever, is a generalization of Mayer's formula for a perfect gas.

To calculate $(\partial c_V/\partial V)_T$, we start from the definition

$$c_V = \left(\frac{\delta Q}{dT}\right)_V = T\left(\frac{\partial S}{\partial T}\right)_V$$

Thus we find by differentiation

$$\left(\frac{\partial c_V}{\partial V}\right)_T = T\frac{\partial^2 S}{\partial V\partial T} = T\frac{\partial^2 S}{\partial T\partial V}$$

Once again we use Maxwell's relation

$$\left(\frac{\partial S}{\partial V}\right)_T = \frac{\partial p}{\partial T}$$

obtaining the result

$$\left(\frac{\partial c_V}{\partial V}\right)_T = T\frac{\partial^2 p}{\partial T^2}$$

The calculation of $(\partial c_p/\partial p)_T$ is analogous. From the definition

$$c_p = \left(\frac{\delta Q}{dT}\right)_p = T\left(\frac{\partial S}{\partial T}\right)_p$$

we find

$$\left(\frac{\partial c_p}{\partial p}\right)_T = T\frac{\partial^2 S}{\partial p\partial T} = T\frac{\partial^2 S}{\partial T\partial p}$$

This time we make use of another of Maxwell's relations

$$\left(\frac{\partial S}{\partial p}\right)_T = -\frac{\partial V}{\partial T}$$

to derive that

$$\boxed{\left(\frac{\partial c_p}{\partial p}\right)_T = -T\frac{\partial^2 V}{\partial T^2}}$$

In brief, if the equation of state of a fluid is known, l and h can be calculated and the other thermal coefficients can be related to a single specific heat, for example c_p. Moreover, thermodynamics allows the derivatives such as $(\partial c_p/\partial p)_T$ to be determined, i.e., the way in which c_p varies with the pressure of a fluid, and therefore its density, at constant temperature can be found. Finally, all the thermal coefficients for all possible states can be obtained if the equation of state is known along with the function $c_p(T, P_0)$ at a pressure P_0 and all temperatures T.

On the other hand, thermodynamics on its own does not enable us to calculate a derivative such as $(\partial c_p/\partial T)_p$ and therefore tells us nothing about $c_p(T, P_0)$ as a function of T. This function must be measured or obtained from a theory that attempts to take account of the microscopic structure of the fluid.

An alternative method

The relations obeyed by the thermal coefficients can be derived without appealing explicit to Maxwell's relations by using a different method. We will describe this method here because it illustrates a very general way of solving certain problems in thermodynamics, namely **dU and dS are considered as exact differentials.**

With the variables T and V for example, we have

$$dU = \delta Q + \delta W = c_V \, dT + (l - p) \, dV$$

and

$$dS = \frac{\delta Q}{T} = \frac{c_V}{T} dT + \frac{l}{T} dV$$

Since dU is an exact differential, we can write

$$\left(\frac{\partial c_V}{\partial V}\right)_T = \left(\frac{\partial (l - p)}{\partial T}\right)_V$$

dS is also an exact differential so that

$$\left(\frac{\partial}{\partial V}\left(\frac{c_V}{T}\right)\right)_T = \left(\frac{\partial}{\partial T}\left(\frac{l}{T}\right)\right)_V = \frac{1}{T}\left(\frac{\partial l}{\partial T}\right)_V - \frac{l}{T^2}$$

91

Thus

$$\left(\frac{\partial c_V}{\partial V}\right)_T = \left(\frac{\partial l}{\partial T}\right)_V - \frac{l}{T}$$

Combining the two expressions for $(\partial c_V/\partial V)_T$, we find one of Clapeyron's relations

$$\left(\frac{\partial p}{\partial T}\right)_V = \frac{l}{T}$$

Substitution of the value obtained for l in this way into one of the expressions for $(\partial c_V/\partial V)_T$ yields

$$\frac{\partial c_V}{\partial V} = T\frac{\partial^2 p}{\partial T^2}$$

From these formulae the others involving $h, c_p - c_V$ and $(\partial c_p/\partial p)_T$ can be derived by appealing to the relations independent of thermodynamics.

4.4 The expansion of a real fluid

In this section we intend to study the temperature changes which occur in a gas when it expands under a variety of conditions without exchanging heat with the surroundings.

JOULE EXPANSION (EXPANSION INTO A VACUUM)

We have already considered this irreversible process in the particular case of a perfect gas. Now we will study the example of a real gas. The gas is allowed to expand from a vessel A into a vessel B which is initially evacuated. The apparatus is assumed to be isolated thermally from its surroundings. The system exchanges neither heat nor work with the surroundings with the result that its internal energy U remains constant.

Let T_1 and V_1 be the temperature and volume before the expansion and T_2 and V_2 the temperature and volume when equilibrium has been established after the expansion.

For a perfect gas, U depends only on T and since $U(T_1) = U(T_2)$ the temperature does not change.

In general for a real gas, U depends on V as well and we must write $U(T_1, V_1) = U(T_2, V_2)$. For U to stay constant, the variation in the volume must be compensated by a variation in temperature. Usually the expansion lowers the temperature. This happens because except when the distance between two molecules is very small, the intermolecular potential (Fig. 1.7) increases with the separation of the molecules. Thus an increase in volume tends to cause an increase in the internal energy except at very high densities. If the internal energy is to stay constant, the temperature must decrease to compensate for this.

The reduction in temperature can be calculated from the condition $U(T_1, V_1) = U(T_2, V_2)$ if the internal energy function $U(T, V)$ is known. For example, for a diatomic gas obeying Van der Waals' equation, we have seen in section 4.2 that

$$U = \tfrac{5}{2}NkT - \frac{A}{V}$$

so that we have

$$T_1 - T_2 = \frac{2A}{5Nk}\left(\frac{1}{V_1} - \frac{1}{V_2}\right)$$

For air at atmospheric pressure expanding into a vacuum and doubling its volume, the reduction in temperature is found to be $0.15°C$.

In the case of an infinitesimal Joule expansion the change of temperature is

$$dT = \left(\frac{\partial T}{\partial V}\right)_U dV$$

$(\partial T/\partial V)_U$ is the *Joule coefficient*. It can be expressed in terms of more familiar quantities by writing $dU = 0$ in terms of the differentials dV and dT

$$0 = dU = T\,dS - p\,dV = T\left[\left(\frac{\partial S}{\partial T}\right)_V dT + \left(\frac{\partial S}{\partial V}\right)_T dV\right] - p\,dV$$

One of Maxwell's relations is

$$\left(\frac{\partial S}{\partial V}\right)_T = \left(\frac{\partial p}{\partial T}\right)_V$$

and we have also for unit mass

$$T\left(\frac{\partial S}{\partial T}\right)_V = \left(\frac{\delta Q}{dT}\right)_V = c_V$$

Thus we find

$$0 = c_V\,dT + \left[T\left(\frac{\partial p}{\partial T}\right)_V - p\right]dV$$

and the Joule coefficient is

$$\left(\frac{\partial T}{\partial V}\right)_U = -\frac{1}{c_V}\left[T\left(\frac{\partial p}{\partial T}\right)_V - p\right]$$

The coefficient is negative, except at very high densities.

The Joule expansion, being an irreversible phenomenon, is necessarily accompanied by an increase in entropy. This increase can be calculated for an infinitesimal expansion by simply writing down the thermodynamic identity which in this example yields

$$0 = dU = T\,dS - p\,dV$$

From this we conclude that

$$\left(\frac{\partial S}{\partial V}\right)_U = \frac{p}{T}$$

This coefficient is always positive as expected.

JOULE–KELVIN EXPANSION

The Joule–Kelvin expansion, another irreversible process, is performed by establishing a permanent steady flow through a porous plug or a throttle (Fig. 4.1). The gas, initially at a pressure p_1 and a temperature T_1, has a pressure p_2 and a temperature T_2 after the expansion. Again it is assumed that there is no exchange of heat with the surroundings.

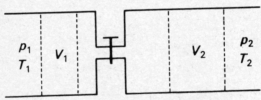

Figure 4.1. A Joule–Kelvin expansion.

Let us follow the displacement of a given mass of gas. It leaves a volume V_1 on one side to occupy a volume V_2 on the other. It receives work p_1V_1 from the pressure on its left-hand surface (Fig. 4.1) and does work p_2V_2 due to the pressure exerted by its right-hand surface. According to the first law, the change in its internal energy will therefore be

$$U_2 - U_1 = p_1V_1 - p_2V_2$$

It follows from this that it is the enthalpy $H = U + pV$ which remains constant during a Joule–Kelvin expansion.

For a perfect gas, since U depends only on the temperature and $pV = NkT$, the enthalpy depends only on the temperature. As $H(T_1) = H(T_2)$, the temperature does not change.

In the general case of a real gas we have

$$H(T_1, P_1) = H(T_2, P_2)$$

94

and if the enthalpy function $H(T, p)$ is known the change in temperature can be calculated.

We will be content here to study the change of temperature for an infinitesimal pressure change by calculating the *Joule–Kelvin coefficient* $(\partial T/\partial p)_H$. Expressing $dH = 0$ in terms of the variations dT and dp, we find

$$0 = dH = T\,dS + V\,dp = T\left[\left(\frac{\partial S}{\partial T}\right)_p dT + \left(\frac{\partial S}{\partial p}\right)_T dp\right] + V\,dp$$

If the equation of state is known, $(\partial S/\partial p)_T$ may be calculated by means of the Maxwell's relation

$$\left(\frac{\partial S}{\partial p}\right)_T = -\left(\frac{\partial V}{\partial T}\right)_p$$

Furthermore

$$T\left(\frac{\partial S}{\partial T}\right)_p = \left(\frac{\delta Q}{dT}\right)_p = c_p$$

for unit mass of gas. Thus

$$0 = dH = c_p\,dT + \left[V - T\left(\frac{\partial V}{\partial T}\right)_p\right]dp$$

and the Joule–Kelvin coefficient is

$$\left(\frac{\partial T}{\partial p}\right)_H = \frac{1}{c_p}\left[T\left(\frac{\partial V}{\partial T}\right)_p - V\right]$$

The sign of the coefficient is determined by the sign of the quantity $T(\partial V/\partial T)_p - V$ which can be obtained from the equation of state. In contrast to the Joule expansion into a vacuum, which nearly always causes cooling, the continuous Joule–Kelvin expansion can produce heating or cooling depending on the initial values of the temperature and pressure. In the (p, T) plane, the loci of the points where $T(\partial V/\partial T) - V$ is zero is a curve, called the inversion curve for the Joule–Kelvin effect; this curve divides the plane into two regions, one where heating occurs, the other where cooling occurs. Figure 4.2 shows the behaviour of nitrogen, a typical example. It can be seen in particular that at low pressure the Joule–Kelvin effect produces heating or cooling depending on whether the initial temperature is above or below the *inversion temperature*, which is about 600 K for nitrogen.

The inversion temperature for hydrogen at low pressure is of the order of 200 K. As a consequence of this, hydrogen heats up when expanded at room temperature; under suitable conditions spontaneous combustion has been observed.

When the initial temperature is below the inversion temperature, a gas is cooled by a Joule–Kelvin expansion. This phenomenon is used with advantage to liquefy certain gases. In particular the Linde machine, which produces liquid air, operates by utilizing the Joule–Kelvin expansion.

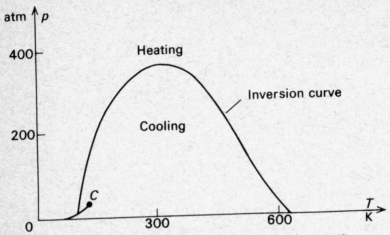

Figure 4.2. The inversion curve for the Joule–Kelvin effect for nitrogen. The liquid–vapour equilibrium curve is also shown. It ends at the critical point C.

Since the Joule–Kelvin expansion is an irreversible phenomenon like the Joule expansion, it is accompanied by an increase in entropy. As the enthalpy is constant, we have

$$0 = \mathrm{d}H = T\,\mathrm{d}S + V\,\mathrm{d}p$$

and so

$$\left(\frac{\partial S}{\partial p}\right)_H = -\frac{V}{T}$$

EXPANSION WITH EXTERNAL WORK

Finally we will consider once more the expansion during which external work is done because the gas pushes back a piston. This process can be considered at least approximately as reversible. We will again discuss the case of an adiabatic expansion in which the gas always cools.

We have already studied the case of a perfect gas for which we have shown that $TV^{\gamma-1}$ is a constant.

In the general case, the fall in temperature is determined by the condition that the entropy is constant during a reversible adiabatic expansion

$$S(T_1, V_1) = S(T_2, V_2)$$

For an infinitesimal expansion we have

$$0 = dS = \left(\frac{\partial S}{\partial T}\right)_V dT + \left(\frac{\partial S}{\partial V}\right)_T dV$$

One of Maxwell's relations is

$$\left(\frac{\partial S}{\partial V}\right)_T = \left(\frac{\partial p}{\partial T}\right)_V$$

and we have also for unit mass

$$\left(\frac{\partial S}{\partial T}\right)_V = \frac{1}{T}\left(\frac{\delta Q}{dT}\right)_V = \frac{c_V}{T}$$

Thus

$$0 = \frac{c_V}{T} dT + \left(\frac{\partial p}{\partial T}\right)_V dV$$

and

$$\left(\frac{\partial T}{\partial V}\right)_S = -\frac{T}{c_V}\left(\frac{\partial p}{\partial T}\right)_V$$

The coefficient is in fact always negative.

Certain gases may be liquefied by means of an expansion which produces external work. George Claude's method of liquefying air is based on this principle. It is a more efficient method of cooling than the irreversible Joule–Kelvin expansion, but the construction of the liquefier requires more sophisticated technology. Since the gas expands against a piston, it is necessary to build a machine which operates at low temperatures and has moving parts.

4.5 The reversible electric cell

Another example of the application of the laws of thermodynamics is provided by the study of a reversible electric cell.

We will consider a cell operating under reversible conditions. The current which it delivers is very small and so the heat produced by the Joule effect inside the cell can be neglected. The state of the cell is defined by the quantity of electricity q that it has delivered and its temperature T. The electromotive force is assumed to be a well-defined function $\varepsilon(q, T)$; ε must not depend on the direction of the current (the cell functions equally well as a generator or a receiver of current) and the cell must be unpolarizable. The Daniell cell and the lead accumulator are examples of cells which cannot be polarized.

When it delivers an infinitesimal quantity of electricity dq, the cell absorbs the electrical work $dW = -\varepsilon\, dq$. In addition, the cell can exchange heat with its surroundings; the infinitesimal heat absorbed has the form $\delta Q = C\, dT + a\, dq$. C is the thermal capacity of the cell and a is a coefficient which expresses the fact that the operation of the cell at constant temperature is in general accompanied by an exchange of heat with the surroundings.[5]

Thermodynamics allows the coefficient a to be related to the variation of ε as a function of T. The problem of establishing this relation is very similar to that of studying the thermal coefficients of a gas. Writing dU and dS as exact differentials, we have

$$dU = \delta W + \delta Q = C\, dT + (a - \varepsilon)\, dq$$

$$dS = \frac{\delta Q}{T} = \frac{C}{T}\, dT + \frac{a}{T}\, dq$$

Thus

$$\left(\frac{\partial C}{\partial q}\right)_T = \left(\frac{\partial (a - \varepsilon)}{\partial T}\right)_p$$

and

$$\frac{1}{T}\left(\frac{\partial C}{\partial q}\right)_T = \left(\frac{\partial}{\partial T}\left(\frac{a}{T}\right)\right)_p = \frac{1}{T}\left(\frac{\partial a}{\partial T}\right)_p - \frac{a}{T^2}$$

Eliminating $(\partial C/\partial q)_T$ from the two equations, the required relation is found

$$a = T\left(\frac{\partial \varepsilon}{\partial T}\right)_p$$

For a Daniell cell, ε in fact depends only on T (and not on q); $d\varepsilon/dT$ is negative, so a is negative and the cell gives out heat when functioning.

4.6 Further remarks on the equilibrium state of a system

The thermodynamic functions make useful tools for investigating the equilibrium of a system under any given conditions. The thermodynamic function to use depends on which variables have fixed values.

---

[5] This is not due to the Joule effect, which is assumed negligible, but to a reversible exchange of heat which can be either an absorption or a production of heat by the cell depending on the sense of the current.

Let us consider first an *isolated system*. It exchanges neither work nor heat with its environment; the variables U and V have fixed values. The thermodynamic function which is suitable for use with the variables (U, V) is the entropy. Regardless of whether the system is in equilibrium or not, there is a particular entropy corresponding to each macroscopic state. We have already seen that *the equilibrium state is the one for which the entropy S is a maximum.*

In the simple case where the deviation of the system from equilibrium can be characterized by an additional variable x, the equilibrium value of x is found by writing

$$\left(\frac{\partial S}{\partial x}\right)_{U,V} = 0$$

for an isolated system.

Let us now consider a system with a fixed volume V in contact with a single source of heat of fixed temperature T, i.e., we no longer have an isolated system. We assume that the system is not in equilibrium and that its state is changing with time. When it passes from one state to another, the change in internal energy ΔU is equal to the heat absorbed Q, since the work done is zero as the volume remains constant. According to the Clausius inequality, the change in entropy is $\Delta S > Q/T$. Thus

$$\Delta(U - TS) = \Delta U - T\Delta S < 0$$

If moreover in its initial and final states, the system has a well-defined temperature T equal to that of the source, $F = U - TS$ is the free energy of the system. Thus we see that *for a system of fixed temperature and volume,* the free energy becomes less as the system develops in time; in other words *the free energy is a minimum in the equilibrium state.*

When the deviation of the system from equilibrium can be characterized by an additional variable x, the equilibrium value of x is found by writing

$$\left(\frac{\partial F}{\partial x}\right)_{T,V} = 0$$

for a system with given T and V.

Finally let us consider a system which can exchange heat and work with a surrounding medium of temperature T and pressure p. When the system, assumed to be not in equilibrium, changes from one state to another, the work and heat absorbed are

$$W = -p\,dV \quad \text{and} \quad Q < T\,dS$$

The change in internal energy is

$$\Delta U = W + Q < T\Delta S - p\Delta V$$

99

and

$$\Delta(U - TS + pV) = \Delta U - T\Delta S + p\Delta V < 0$$

If in addition the system in its initial and final states has a well-defined temperature T and pressure p equal to that of the external medium, $G = U - TS + pV$ is the thermodynamic potential or Gibbs free energy. Thus we see that *for a system with a given temperature and pressure*, the thermodynamic potential becomes less as the system changes in time; in other words *the thermodynamic potential is a minimum in the equilibrium state.*

In the simple case in which the deviation from equilibrium can be characterized by an additional variable x, the equilibrium value of x can be obtained from

$$\left(\frac{\partial G}{\partial x}\right)_{T,p} = 0$$

for a system with a given T and p.

To sum up, we find that for each pair of fixed variables $(U, V), (T, V), (T, p)$, the appropriate thermodynamic function $S(U, V)$, $F(T, V)$, $G(T, p)$ has a minimum value when a system is in equilibrium.

The case where the specified variables are (T, p) is of particular practical importance because it applies to the common case of a system maintained at room temperature and atmospheric pressure.

4.7 Phase changes

In this section we are going to make use of thermodynamics to increase our understanding of phase changes in a pure substance.

EQUILIBRIUM BETWEEN PHASES

We have already seen that, depending on the conditions, the equilibrium state of a pure substance consists of a single phase or possibly two or even three phases in coexistence. These various possibilities follow from the properties of the thermodynamic potential $G(T, p)$.

For a fixed temperature T and pressure p, let us consider two possibilities 1 and 2 for the phase in which the substance occurs (solid or liquid for example). The thermodynamic potentials $G_1(T, p)$ and $G_2(T, p)$ which correspond to these possibilities are generally different. The stable phase is the one for which the thermodynamic potential is lower. The other thermodynamic potential can thus be interpreted as that of a metastable state.

However it can happen that G_1 and G_2 are equal; in the (p, T) plane, the equation $G_1(p, T) = G_2(p, T)$ determines the equilibrium curve between the two phases. In fact on the curve for which $G_1 = G_2$, a system consisting of a fraction x by weight of phase 1 and a fraction $(1 - x)$ of phase 2 will have a thermodynamic potential

$$xG_1 + (1 - x)G_2 = G_1 = G_2$$

which is independent of x; the two phases can coexist in equilibrium in any proportions whatsoever. In the representation in the (p, V) plane, the isobaric transition region represents all the states where the two phases coexist with the same values of p and T.

If G_1, G_2, G_3 are the thermodynamic potentials of the solid, liquid, and gaseous phases, the coordinates (p, T) of the triple point are determined by the equations $G_1(p, T) = G_2(p, T) = G_3(p, T)$.

The fact that the thermodynamic potentials for the liquid and gaseous phases under temperature and pressure conditions corresponding to possible equilibrium between these phases are equal enables us to verify that *Maxwell's construction* is valid. It should be recalled that an approximate theory, such as Van der Waals', which does not take account of the possibility of the existence of two phases, can produce an isothermal $p(V)$ with an oscillating region. Using Maxwell's construction, this region is replaced by an isobaric region, positioned so that the shaded areas in Fig. 1.9 are equal.

Although the dashed part of the isothermal represents states with a single phase which are not equilibrium states (the stable states are on the isobaric part of the curve), we will suppose that the fluid can be taken from state A to state B following the dashed curve. Along this isothermal curve the temperature is constant and the differential of the thermodynamic potential $dG = -S\,dT + V\,dp$ reduces to $dG = V\,dp$. Between the points A and B the total change in G is zero, since the pure vapour at A and the pure liquid at B have the same thermodynamic potential. Thus

$$\int_A^B V\,dp = 0$$

where the integral is taken along the dashed curve. The geometric interpretation of this integral shows that the shaded areas are equal as required for Maxwell's construction.

The same result may be obtained by different reasoning. Let us consider the cycle A–dashed curve–B–isobaric region–A. As the temperature is the same throughout the cycle, it follows from one of the versions of the second law that the work done by the fluid, $\oint p\,dV$, can only be negative or zero. If we go round the cycle in the opposite direction, the magnitude of the work done is the same but the sign is changed. Since the work done must again be negative or zero, the only solution is that the work done is zero and $\oint p\,dV = 0$. Thus the two shaded areas must be equal.

When a substance changes phase, it absorbs or produces heat.

For a solid to melt or sublime or a liquid to evaporate, it is necessary to supply a quantity of heat L per unit mass, called *the latent heat* of fusion, sublimation, or vaporization respectively. This definition of latent heat assumes that the pressure is constant. The temperature of course remains constant throughout a change of phase. The latent heat supplies the extra internal energy required for the rearrangement of the atoms during the phase change.

Clapeyron's formula. Thermodynamics enables an important relation concerning the latent heat to be established. Let us consider for example a liquid–vapour phase change occurring at a point in the (p, V) plane on the isobaric region of liquefaction of the isotherm T (Fig. 4.3). We will work with unit mass of the substance.

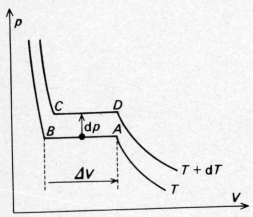

Figure 4.3. Isobaric liquefaction regions in the (p, V) plane.

One of Maxwell's relations is

$$\left(\frac{\partial S}{\partial V}\right)_T = \left(\frac{\partial p}{\partial T}\right)_V$$

$(\partial S/\partial V)_T$ is the derivative of the entropy taken along the isobaric region. Since $dS = \delta Q/T$ and the heat absorbed is a linear function of the mass involved and therefore of the change in volume, we have

$$\left(\frac{\partial S}{\partial V}\right)_T = \frac{1}{T}\frac{L}{\Delta V}$$

102

where ΔV is the difference between the volume of unit mass of the vapour and of the liquid. If we increase T by dT at constant V, we transfer to the isobaric part of the isotherm for $T + dT$; $(\partial p/\partial T)_V$ is therefore simply the derivative dp/dT of the saturated vapour pressure $p(T)$ of Fig. 1.14. Hence *Clapeyron's formula* is obtained

$$\frac{dp}{dT} = \frac{1}{T}\frac{L}{\Delta V}$$

It should be noted that this formula of Clapeyron concerning a change of phase is closely related to his relation for a homogeneous fluid, $l = (\delta Q/dV)_T = T(\partial p/\partial T)$; we have proved both relations in essentially the same way.

Another possible proof consists of considering the journey round the infinitesimal cycle $ABCD$ (Fig. 4.3). For this cycle, the work produced, $\oint p\, dV$, is the area $\Delta V\, dp$. The heat absorbed from the hot source at $T + dT$ is infinitely close to L. According to Carnot's theorem, the efficiency is

$$\frac{\Delta V\, dp}{L} = \frac{dT}{T}$$

Once again we have found Clapeyron's formula.

Here we have discussed vaporization but Clapeyron's relation is of course equally applicable to sublimation and fusion.

It follows from Clapeyron's formula that the sign of dp/dT is that of $L/\Delta V$. For vaporization and sublimation L and ΔV are always positive and the equilibrium curve $p(T)$ has in fact a positive slope.

On the other hand for fusion, L is positive[6] but the sign of ΔV depends on the substance. Generally, ΔV is positive for fusion and so dp/dT is positive. However, in certain cases, notably that of water, ΔV is negative (there is a decrease in volume when ice melts). Thus the $p(T)$ curve has a negative slope; ice can be made to melt by compressing it.

SATURATED VAPOUR PRESSURE

For vaporization and sublimation, an approximate formula for the saturated vapour pressure can be derived from the exact relation of Clapeyron.

We make the approximations that the latent heat L is a constant, neglecting any variation of L along the equilibrium curve $p(T)$, and also that ΔV is the volume V of the gaseous state, neglecting the volume of the condensed state. Then we apply the perfect gas law $pV = NkT$ to the vapour and obtain

$$\frac{dp}{dT} = \frac{L}{T}\frac{p}{NkT}$$

[6] Helium 3 is exceptional. Below 0·3 K, heat must be supplied to *solidify* the liquid.

This can be integrated immediately, yielding

$$p = p_0 \exp(-L/NkT)$$

where p_0 is the constant of integration. The latent heat is usually defined for unit mass so N must be taken as the number of molecules of the vapour contained in unit mass. It follows from this exponential law that the saturated vapour pressure decreases very quickly as the temperature falls.

Very low temperatures between 1 and 5 K can be measured utilizing a somewhat improved version of the formula above. The saturated vapour pressure of helium is measured and the absolute temperature deduced from it.

Problems

4.1 Ice has a latent heat of fusion of $3.34 \times 10^5 \, \mathrm{J \, kg^{-1}}$ and its density is $917 \, \mathrm{kg \, m^{-3}}$. What is the slope of the fusion curve at $0°C$?

What pressure is required to melt ice maintained at $-2°C$?

4.2 The isothermal coefficient of compressibility of ice is

$$\chi = -\frac{1}{V}\left(\frac{\partial V}{\partial p}\right)_T = 1.2 \times 10^{-10} \, \mathrm{m^2 \, N^{-1}}$$

and its isobaric coefficient of expansion is

$$\alpha = \frac{1}{V}\left(\frac{\partial V}{\partial T}\right)_p = 1.57 \times 10^{-4} \, \mathrm{deg^{-1}}$$

A block of ice at $-2°C$ under atmospheric pressure is heated at constant volume. Use the results of problem 4.1 to determine the temperature at which it will melt.

4.3 Show that the slope of the sublimation curve at the triple point is greater than that of the vaporization curve.

4.4 Show that for any fluid whatsoever, the specific heats at constant volume, c_V, and constant pressure, c_p, are connected by the relation

$$c_p - c_V = \frac{\alpha^2 T V}{\chi}$$

where

$$\alpha = \frac{1}{V}\left(\frac{\partial V}{\partial T}\right)_p; \; \chi = -\frac{1}{V}\left(\frac{\partial V}{\partial p}\right)_T$$

The volume V and the temperature T should be used as the independent variables.

4.5 Express the heat absorbed by any fluid in a reversible transformation as a function of c_V, c_p, α, χ, T and V defined as in problem 4.4.

The coefficient of expansion of water has the value 15×10^{-6} deg^{-1} at 5°C. What is the final temperature of water that is compressed isentropically from 1 to 3 000 atm?

4.6 The latent heat of vaporization of water at 100°C is $2 \cdot 25 \times 10^6$ J kg^{-1}. Assuming that the saturated vapour pressure of water in the region of 100°C follows a law of the form

$$p = \left(\frac{t°C}{100}\right)^{\alpha}$$

where p is expressed in atmospheres, calculate α.

4.7 Show that for an isothermal transformation, in which T is no longer considered as an independent variable, the amount of work absorbed, dW, is the differential of a function of the other independent variables describing the transformation. Repeat the problem for an isentropic transformation.

4.8 A piezoelectric quartz crystal consists of a sheet of surface area S and thickness e placed between two metal plates forming a parallel plate condenser with charge q and potential difference V. A pressure variation dN applied to the sheet causes a variation of its thickness de.

(a) If the thickness is varied in an adiabatic and reversible manner keeping the potential difference V a constant, the change in charge dq is αde. When the potential difference is varied by dV in an adiabatic way while the thickness is kept constant, show that the variation in the pressure produced is

$$dN = -\frac{\alpha \, dV}{S}$$

(b) Repeat the problem above assuming the transformations are isothermal (use the property of work established in problem 4.7).

(c) With $q = 0$, what potential difference is produced when a pressure is applied to the sheet in an adiabatic and reversible way to produce a change in thickness de?

4.9 A tension τ is applied slowly and adiabatically to a metallic wire of length L and temperature T. What is the temperature of the wire when the tension τ is reached?

If the wire has an isothermal coefficient of elasticity

$$\alpha = \frac{1}{L}\left(\frac{\partial L}{\partial \tau}\right)_T$$

what is the temperature of the wire after the tension has been released?

4.10 A metallic wire experiences a couple C which produces a twist θ such that $C = \gamma\theta$.

(a) If the couple is applied in an adiabatic and reversible manner, what is the resulting temperature of the wire?

(b) If the couple is suddenly removed, what is the new temperature?

4.11 A real gas is expanded in an adiabatic, irreversible way from a pressure p_1 and a temperature T_1 to a pressure p_2. Assuming its specific heat at constant volume is a constant, determine the final temperature for

(a) a perfect gas,

(b) a gas obeying the equation of state $p(V - b) = rT$,

(c) a gas obeying Van der Waals' equation of state

$$\left(p + \frac{a}{V^2}\right)(V - b) = rT$$

4.12 If a surface separating two phases, for example a soap bubble, is stretched, an amount of work $\delta W = A\,dS$, where A is the surface tension, must be supplied.

Show that the change in internal energy of the surface is

$$dU = dS\left(1 - T\frac{\partial A}{\partial T}\right)$$

4.13 A Daniell cell has an e.m.f. of 1·0934 V at 273 K. This e.m.f. decreases by 45×10^{-5} V deg^{-1}. What is the change in the internal energy of the cell when it supplies 2 F ($2 \times 96\,500$ C) isothermally? If the cell supplies this charge adiabatically, what change of temperature results? Assume that the cell has a thermal capacity of $1\,000$ J deg^{-1}.

5. The Boltzmann distribution and some of its applications

This chapter is devoted to the study of systems whose energy fluctuates because they exchange energy with their surroundings. Under fixed conditions, the probability that the system has a particular energy can be determined by means of statistical mechanics. We are going to discuss the law governing this probability and describe some applications of it.

We will begin by studying a simple example from a macroscopic point of view.

5.1 Equilibrium of an isothermal atmosphere in a gravitational field

In this section we intend to investigate how the density of a perfect gas at a uniform temperature varies with altitude under the action of gravity.

Let us consider a volume of gas in the form of a sheet of unit area and thickness dz (Fig. 5.1). Let m be the mass of each molecule, n the number of

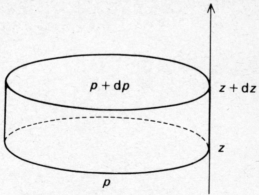

Figure 5.1. A sheet of gas in equilibrium.

molecules per unit volume at an altitude z and g the acceleration due to gravity. We choose the direction of the z axis to be upwards.

The weight of the sheet of gas is $-mgn \, dz$. If p and $p + dp$ are the pressures at heights z and $z + dz$ respectively, then the net force on the sheet due to the pressure is $-dp$. At equilibrium

$$-mgn \, dz - dp = 0$$

The perfect gas law may be written as $p = nkT$, and hence we can write

$$mgn \, dz = -kT \, dn$$

This relation between n and z can be integrated to give

$$n = n_0 \, e^{-mgz/kT}$$

where n_0 is the constant of integration.

Thus the density n decreases exponentially as the altitude z increases. It should be noticed that this law only applies if the temperature is uniform. Since the temperature of the Earth's atmosphere is not uniform, the exponential law is only an approximation.

This law for the variation of n may also be expressed in the form

$$n = n_0 \, e^{-E/kT}$$

where $E = mgz$ is the potential energy of a molecule at height z. Some important generalizations can be made from this formula.

5.2 A general treatment of the Boltzmann distribution

In the example we have just considered, we saw that the density of particles with an energy E is proportional to $e^{-E/kT}$. This result is extremely general as we will now show.

Let us consider any small system Φ (microscopic or macroscopic) which is completely surrounded by a much larger system Φ', (Fig. 5.2). Φ' is therm-

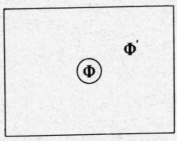

Figure 5.2. A small system Φ in thermal contact with a large system Φ'.

ally isolated and the two systems together have an energy E_0 which has a well-defined value (to within the precision of measurement). The interaction energy between Φ and Φ' is assumed to be sufficiently small so that E_0 can be considered as the sum of the energy E of the small system and the energy $E_0 - E$ of the large. We intend to calculate the probability $P(E)$ of the small system occupying a microscopic state α of energy E when equilibrium is reached.

When the small system has energy E, the energy of the large system is $E_0 - E$; let $\Omega'(E_0 - E)$ be the number of microscopic states of this energy for the large system. $P(E)$ is the probability that Φ is found in state α and Φ' in any of the Ω' states accessible to it. The number of accessible states of the complete system $\Phi + \Phi'$ is therefore $1 \times \Omega' = \Omega'$. $P(E)$ is proportional to this number:

$$P(E) = C\Omega'(E_0 - E)$$

where C is a constant.

$\Omega'(E_0 - E)$ varies extremely rapidly as a function of E and a more easily handled function is obtained by taking logarithms:

$$\log P(E) = \log C + \log \Omega'(E_0 - E)$$

By definition, $\log \Omega'$ is the entropy S' of the large system, Φ', to within a factor k and thus we may write

$$\log P(E) = \log C + \frac{1}{k} S'(E_0 - E)$$

As E is small compared with E_0, the entropy S' can be expanded as a Taylor series and terms above second order neglected:

$$\log P(E) = \log C + \frac{1}{k} S'(E_0) - \frac{1}{k} \frac{dS'}{dE_0} E$$

dS'/dE_0 is by definition the reciprocal $1/T$ of the temperature of the large system. This temperature is not changed appreciably by small fluctuations in the energy of the large system; there is essentially no difference between the derivative of S' for the energy E_0 and that for the exact energy $E_0 - E$ of the large system. $\log P$ finally reduces to the form

$$\log P(E) = B - \frac{E}{kT}$$

where B is a constant. From this we find that

$$P(E) = A\,e^{-E/kT}$$

where A is a constant.

This result expresses the desired generalization which can be formally stated as follows: if a system Φ is in contact with a system much larger than itself at a temperature T, the probability of the system Φ occupying a microscopic state of energy E is proportional to the *Boltzmann factor* $e^{-E/kT}$.

The constant A is determined from the normalization condition that the sum of $P(E)$ for all possible states must be equal to unity.

One very important remark should be made. The Boltzmann factor gives the probability of the system being in *one* microscopic state of energy E. For a small region close to E containing $d\omega$ states the probability of the system being in *any* state is

$$P(E)\, d\omega = A\, e^{-E/kT}\, d\omega$$

MICROSCOPIC AND MACROSCOPIC SYSTEMS

The Boltzmann formula is valid if the system Φ is microscopic or macroscopic. Its consequences are qualitatively different in the two cases.

For a microscopic system the states can be considered one by one and the probability of the system occupying a particular state investigated. The Boltzmann formula gives this probability.

The energy of a microscopic system fluctuates in a way which can be important on the microscopic scale of the system.

For a macroscopic system, the microscopic states have an extremely complex structure and it is meaningless to investigate a particular microscopic state. What we want to know is the probability of the system having a certain energy. If $\Omega(E)$ is the (very large) number of microscopic states which have energy E (to within the precision of the measurements), the probability of the system being in any one of the microscopic states of energy E is

$$P(E)\Omega(E) = A\, e^{-E/kT}\Omega(E)$$

As we have seen in section 3.5, the function $\Omega(E)$ increases extremely rapidly with E. The exponential $P(E)$ is a decreasing function of E and so the product $P(E)\Omega(E)$ has a very sharp maximum for some value U of the energy. As a result of this, the system almost certainly has the energy U. The energy fluctuations of a macroscopic system in contact with a constant temperature source are negligible on a macroscopic scale. There is essentially no difference between an isolated macroscopic system and the same system in equilibrium with such a source. In both cases the internal energy U is well-defined.

5.3 Paramagnetism

As our first example of the application of the Boltzmann distribution, we will study paramagnetism.

MAGNETIZATION OF A PARAMAGNETIC SUBSTANCE

Substances which are not spontaneously magnetic become weakly magnetized when placed in an external magnetic field. If the induced magnetic moment is in the opposite direction to the field, the substance is said to be diamagnetic. If the moment is in the same direction, the substance is said to be *paramagnetic*. It is the latter that we intend to consider in this section.

A paramagnetic substance consists of atoms which carry a permanent magnetic moment. If there is no external field the atoms are oriented randomly and the resulting total magnetic moment is zero. However, an external field tends to orient the magnetic moments parallel to itself and a non-zero total magnetic moment is produced. It is the non-zero moment which we are going to calculate.

We will consider the simple case, already studied in connection with the concept of entropy, in which the atoms have a spin $\frac{1}{2}$ and a magnetic moment μ.

If the atoms are placed in a uniform magnetic field of induction $\mathbf{B}$, the forces exerted by the field on a magnetic moment $\boldsymbol{\mu}$ produce a potential energy[1] $-\boldsymbol{\mu} \cdot \mathbf{B}$. If the direction of $\mathbf{B}$ is taken as the axis of quantization, the projection of the magnetic moment of an atom on this axis can take the two values $+\mu$ and $-\mu$ and so the energy of the atom is $-\mu B$ or $+\mu B$.

Each atom constitutes a system with two possible states, which is in contact with the rest of the substance at temperature T. According to Boltzmann's law, the probabilities of the two states are $A \, \mathrm{e}^{\mu B/kT}$ and $A \, \mathrm{e}^{-\mu B/kT}$ respectively. The normalization condition requires that

$$A = \frac{1}{\mathrm{e}^{\mu B/kT} + \mathrm{e}^{-\mu B/kT}}$$

The mean value of the projection of the magnetic moment of an atom is thus (see Appendix II)

$$\mu A \, \mathrm{e}^{\mu B/kT} - \mu A \, \mathrm{e}^{-\mu B/kT} = \mu \tanh \frac{\mu B}{kT}$$

If there are N/V magnetic atoms per unit volume, the magnetic moment per unit volume, or the magnetization, will be

$$\mathscr{J} = \frac{N}{V} \mu \tanh \frac{\mu B}{kT}$$

The variation of $\mathscr{J}$ as a function of B has the form shown in Fig. 5.3. Qualitatively, analogous results are found for spins other than $\frac{1}{2}$.

[1] Some authors refer to this as mechanical energy. It must not be confused with the total electromagnetic energy; in calculating the total energy it is necessary to account for the energy required to create the field itself.

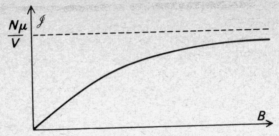

Figure 5.3. The magnetization $\mathscr{I}$ as a function of the applied field B.

At very large values of B, $\mathscr{I}$ tends to saturate at a value $N\mu/V$; the atoms are then completely aligned. For a finite value of B, $\mathscr{I}$ is determined by a compromise between the action of the field, which tends to align the atoms, and thermal agitation, which tries to orient them randomly.

If B is sufficiently weak, or more precisely if μB is small compared with kT, $\tanh(\mu B/kT)$ may be replaced by $\mu B/kT$ and the magnetization is then proportional to the field

$$\mathscr{I} = \frac{N\mu^2}{VkT} B$$

In the SI system of units the relation $\chi = \mu_0 \mathscr{I}/B$, where μ_0 is the permeability of free space, calculated in the weak field limit is a dimensionless quantity called the *magnetic susceptibility*. In this case we have

$$\chi = \mu_0 \frac{N\mu^2}{VkT}$$

Thus the magnetic susceptibility of a paramagnetic substance is inversely proportional to the absolute temperature. This is known as *Curie's law*.

Curie's law can be applied to measure very low temperatures below 1 K. The magnetic susceptibility of a paramagnetic substance is measured and the absolute temperature deduced from it.

ADIABATIC DEMAGNETIZATION

In section 3.2 we calculated the entropy of a system of N spins of $\frac{1}{2}$ when $(N + n)/2$ spins were aligned in one direction and $(N - n)/2$ spins were aligned in the opposite direction. In the present discussion these directions are chosen to be along the direction of the field **B**. We have shown

$$S = k \log \frac{N!}{\left(\dfrac{N + n}{2}\right)! \left(\dfrac{N - n}{2}\right)!}$$

Thus S depends only on the number of atoms aligned in each direction. These numbers are proportional to the respective Boltzmann factors $e^{\mu B/kT}$ and $e^{-\mu B/kT}$, which depend only on the ratio B/T. Therefore we see that the entropy depends not on B or T separately, but only on the *ratio* B/T. This result holds for spins other than $\frac{1}{2}$.

At low temperatures, the entropy of a paramagnetic substance arises mainly from the entropy of the spins, since the thermal agitation of the atoms and the corresponding entropy is very small. We have just seen that this entropy has the form $S(B/T)$. Let us suppose the applied field **B** is varied in an adiabatic reversible manner, i.e., the substance must be thermally isolated and, as is the case under the usual experimental conditions, the field must vary slowly compared with the time required for equilibrium to be established. The entropy remains unchanged as we saw in section 3.3. Here the constant entropy is particularly easy to understand since a 'slow' change of the field does not alter the distribution of magnetic moments parallel and antiparallel to the field. As the function $S(B/T)$ does not change, the ratio B/T remains constant and the absolute temperature varies proportionally to the field.

If the field B is reduced, the temperature decreases as well. We have at our disposal here an extremely powerful method of producing very low temperatures. A paramagnetic substance in contact with liquid helium at a temperature of the order of 1 K is first subjected to an intense magnetic field of the order of several teslas and then thermally isolated. If the field is now reduced, the substance cools. This process is called *adiabatic demagnetization*.

The temperature cannot however be reduced indefinitely by this method. We have implicitly assumed the paramagnetic material is perfect, i.e., we have neglected the interactions between the magnetic moments of the atoms. In practice these interactions are present and even when the applied field has been reduced to zero, each atom experiences a net field due to its neighbour. Thus the total field can never be reduced to zero as is required to attain $T = 0$. With ordinary paramagnetic substances, temperatures of the order of 10^{-3} K can be reached.

In a normal paramagnetic material, the magnetic moments of the atoms arise from the electrons. Nuclear paramagnetism is exhibited by materials in which the nuclei of the atoms carry a magnetic moment. These nuclear moments are very much smaller and the interactions between them very much weaker. Adiabatic demagnetization of a nuclear paramagnetic material enables even lower temperatures of the order of 10^{-6} K to be reached.

NEGATIVE ABSOLUTE TEMPERATURES

With some nuclear paramagnetic materials, equilibrium states can be achieved which have the curious property of corresponding to a negative absolute temperature.

We will consider the simple case of a nucleus of spin $\frac{1}{2}$. The specimen, subjected to a field **B**, is in equilibrium with its surroundings at the temperature T (positive). According to Boltzmann's law we have

$$P(E) = A\,e^{-E/kT}$$

and so the nuclei whose magnetic moments are aligned in the direction of **B** and which have a negative energy $E = -\mathbf{\mu}\,.\,\mathbf{B}$, are more numerous than those aligned in the opposite direction.

If the field **B** is suddenly reversed, the magnetic moments have no time to reverse. Their energies change sign so that the magnetic moments with positive energy are now more numerous. The distribution is now given by

$$P'(E) = A\,e^{+E/kT}$$

which describes a system of spins in equilibrium with each other at a temperature $-T$.

Such an experiment is only possible because the specimen employed contains nuclear magnetic moments which interact very little with their surroundings, and take a long time to reach equilibrium with it. After a sufficiently long time (several minutes) the nuclear spin system regains the positive temperature of the surroundings.

In section 3.5 we showed that the absolute temperature is always positive for systems whose entropy is an increasing function of the internal energy. This is in fact generally the case. The behaviour of the entropy of a system of spins in a magnetic field is exceptional. The internal energy U has an upper limit corresponding to the perfectly ordered state in which all the magnetic moments are aligned in the opposite direction to the field. When U approaches this limit, the entropy S becomes a decreasing function of U and $1/T = \partial S/\partial U$ is then negative.

5.4 The Maxwell distribution

Let us return to the investigation of the velocities of the molecules in a gas. In section 1.4 we discussed only the *mean* translational kinetic energy per molecule which is

$$\tfrac{1}{2}m\overline{v^2} = \tfrac{3}{2}kT$$

Therefore at a given temperature we know $\overline{v^2}$. This does not mean that all the molecules have the speed $\sqrt{3kT/m}$; some travel faster and others slower so that $\overline{v^2}$ is merely the mean square speed. We may write

$$\overline{v^2} = \frac{1}{N}\sum_{i=1}^{N}\overline{v_i^2}$$

where v_i is the speed of the ith molecule.

114

It is convenient to represent a velocity **v** by a point in a three-dimensional space, called velocity space; the point is specified by the cartesian coordinates v_x, v_y, and v_z. We cannot expect to catalogue all the individual velocities v_i; their number is of the order of 10^{23}. However, we can calculate a simpler quantity, namely the fraction of the total number of molecules whose velocity is represented by a point inside the volume element $d^3\mathbf{v} = dv_x\, dv_y\, dv_z$ around the point **v** in velocity space.

We denote this fraction, which is obviously proportional to $d^3\mathbf{v}$, by the expression $f(\mathbf{v})\, d^3\mathbf{v}$; $Nf(\mathbf{v})\, d^3\mathbf{v}$ is then the number of molecules whose velocity is represented by a point in $d^3\mathbf{v}$. In other words, $f(\mathbf{v})\, d^3\mathbf{v}$ is the probability that a molecule has a velocity represented by such a point. Although the individual velocities vary with time, the function $f(\mathbf{v})$ itself does not change once equilibrium has been reached. The function $f(\mathbf{v})$ is called the *Maxwell distribution*.

We will determine the function $f(\mathbf{v})$ by the application of Boltzmann's formula. Classical mechanics will be adequate in this calculation. The number $d\omega$ of velocity states for a molecule whose velocity is in $d^3\mathbf{v}$ must be proportional to $d^3\mathbf{v}$ (we have enumerated the velocity states using a rule analogous to the one we employed in section 3.4 for the position states, namely that the number of such states in a certain volume is proportional to that volume). The energy of a molecule with speed v is $E = \frac{1}{2}mv^2$. According to Boltzmann's formula, the probability of a molecule being in one of the velocity states in $d^3\mathbf{v}$ is then

$$P(E)\, d\omega = A\, e^{-E/kT}\, d^3\mathbf{v} = A\, e^{-mv^2/2kT}\, d^3\mathbf{v}$$

where A is a constant. Thus the Maxwell distribution is simply the Boltzmann factor

$$f(v) = A\, e^{-mv^2/2kT}$$

It should be noticed that f depends only on the modulus v of the velocity; the molecules are distributed isotropically in velocity space.

The constant A can be determined from the normalization condition that the total probability must be unity.[2]

$$1 = \iiint_{-\infty}^{\infty} f(v)\, dv_x\, dv_y\, dv_z$$

$$= \iiint_{-\infty}^{\infty} A\, e^{-(m/2kT)(v_x^2 + v_y^2 + v_z^2)}\, dv_x\, dv_y\, dv_z$$

$$= A\left(\frac{2\pi kT}{m}\right)^{\frac{3}{2}}$$

[2] The calculus of this type of integral is outlined in Appendix III.

The complete expression for the Maxwell distribution can be obtained by substituting for A

$$f(v) = \left(\frac{m}{2\pi kT}\right)^{\frac{3}{2}} e^{-mv^2/2kT}$$

The molecular interactions depend on the positions of the molecules but not on their speeds. In consequence the Maxwell distribution is valid whatever interactions are present and applies not only to perfect gases but also to real fluids and even solids, provided only that the laws of classical mechanics hold.

It is worth stressing the fact that the probability of finding a molecule with a velocity of exactly $\mathbf{v}$ is zero, just as the probability of finding a molecule precisely at a specified point is zero. A non-zero result is obtained only for the probability of finding the velocity in a particular element of velocity space or the position in a particular element of ordinary space. $f\,d^3\mathbf{v}$ is a probability; f itself is only a probability density.

$f(v)\,d^3\mathbf{v}$ is the probability that, to within $d\mathbf{v}$, a molecule has a velocity $\mathbf{v}$; both the *magnitude* and the *direction* of the velocity are specified. This probability is a maximum when $v = 0$ and decreases as v increases. Another interesting quantity is the probability $F(v)\,dv$ that the modulus of the velocity of a molecule lies between v and $v + dv$. $F(v)\,dv$ is obtained by summing $f(v)\,d^3v$ over all points in velocity space whose distance from the origin lies between v and $v + dv$. The corresponding volume in velocity space is the volume of a thin spherical shell of thickness dv and radius v namely $4\pi v^2\,dv$. Thus

$$F(v)\,dv = f(v)4\pi v^2\,dv = 4\pi\left(\frac{m}{2\pi kT}\right)^{\frac{3}{2}} v^2\, e^{-mv^2/2kT}\,dv$$

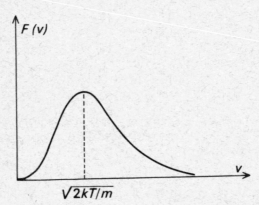

Figure 5.4. The distribution $F(v)$ of the modulus of the velocity as a function of v.

116

The function $F(v)$ has the form $v^2\,e^{-mv^2/2kT}$ and is shown in Fig. 5.4. This function has a maximum when the derivative with respect to v is zero:

$$\frac{d}{dv}(v^2\,e^{-mv^2/kT}) = 0$$

The corresponding value of v is

$$v = \sqrt{2kT/m}$$

and this is the *most probable speed*.

The mean square speed can be shown to be

$$\overline{v^2} = \int_0^\infty v^2 F(v)\,dv = \frac{3kT}{m}$$

in agreement with the elementary definition of temperature in the formula $\frac{1}{2}mv^2 = \frac{3}{2}kT$.

5.5 The equipartition of energy in classical statistical mechanics

The contribution of a single cartesian component v_x of the velocity of a molecule to the kinetic energy is $\frac{1}{2}mv_x^2$. This energy has the mean value

$$\tfrac{1}{2}m\overline{v_x^2} = \tfrac{1}{2}kT$$

We will show how this result can be generalized.

Let us consider any macroscopic system whose energy naturally depends on all the microscopic variables characterizing it such as the positions and speeds of the molecules. Choosing one of these variables q, we assume that the energy has the form

$$E = aq^2 + \ldots$$

where a is a constant and the other terms in the expression depend on variables other than q. v_x is a particular example since

$$E = \tfrac{1}{2}mv_x^2 + \ldots$$

We wish to determine the mean value of aq^2 when the system is in equilibrium. We assume that the conditions are such that the laws of classical mechanics can be applied to the variable q.

The problem is formally the same as that of determining the mean value of the energy aq^2 of a microscopic system which can exchange energy with the rest of the macroscopic system of which it is a part. The Boltzmann formula can again be applied classically to calculate the possible states. The probability of the value of q lying in an interval dq is

$$A\,e^{-aq^2/kT}\,dq$$

The mean value of aq^2 is

$$\overline{aq^2} = A \int_{-\infty}^{\infty} aq^2\,e^{-aq^2/kT}\,dq$$

The constant A is determined by the normalization condition for the probability

$$A \int_{-\infty}^{\infty} e^{-aq^2/kT}\,dq = 1$$

Thus we have

$$\overline{aq^2} = \frac{\int_{-\infty}^{\infty} aq^2\,e^{-aq^2/kT}\,dq}{\int_{-\infty}^{\infty} e^{-aq^2/kT}\,dq}$$

The integrations may be avoided by proceeding as follows. We put $1/kT = \beta$ and note that

$$\overline{aq^2} = -\frac{\partial}{\partial \beta} \log\left(\int_{-\infty}^{\infty} e^{-\beta aq^2}\,dq\right)$$

Taking the variable of integration as $y = (\beta a)^{\frac{1}{2}}q$, we obtain

$$\overline{aq^2} = -\frac{\partial}{\partial \beta} \log\left[(\beta a)^{-\frac{1}{2}} \int_{-\infty}^{\infty} e^{-y^2}\,dy\right] = \frac{\partial}{\partial \beta}(\tfrac{1}{2}\log \beta) + \cdots$$

where the terms which have been ignored in the final expression do not depend on β and so do not contribute to the derivative with respect to β. Finally we obtain

$$\overline{aq^2} = \frac{1}{2\beta} = \frac{kT}{2}$$

This result expresses the required generalization which is known as the principle of the *equipartition of energy*. This states that when classical mechanics can be applied the mean value of any quadratic term of the form aq^2 in the expression for the energy of the system is $\tfrac{1}{2}kT$.

5.6 Specific heats of solids

CLASSICAL THEORY

In a solid, thermal agitation causes the atoms to oscillate about their mean positions. To a first approximation, we can assume that each atom is subjected to a restoring force proportional to its displacement. The atom behaves as a three-dimensional harmonic oscillator. If the cartesian components of the displacement of an atom are denoted by x, y, and z and the velocity components by v_x, v_y, and v_z, its energy is

$$\tfrac{1}{2}m(v_x^2 + v_y^2 + v_z^2) + \tfrac{1}{2}K(x^2 + y^2 + z^2)$$

where K is the spring constant.

The energy depends on six variables and so the atom is said to have six degrees of freedom. Since the energy is a quadratic function of these variables, the principle of the equipartition of energy can be applied and the mean value of each energy term is $\tfrac{1}{2}kT$.

The total energy of the atom therefore has the mean value $3kT$. If a solid formed from N atoms is considered, the internal energy must be $U = 3NkT$, and so the thermal capacity is

$$C_V = \frac{dU}{dT} = 3Nk$$

For a gram-atom of the solid, N is Avogadro's number and the thermal capacity is

$$C_V = 3R \sim 25 \, \text{J K}^{-1} \sim 6 \, \text{cal K}^{-1}$$

At ordinary temperatures the experimental value of thermal capacity per gram-atom at constant volume[3] for most solids does in fact agree with this result. The relation was discovered empirically by *Dulong and Petit*. As we have seen above, it is explained by classical statistical mechanics.

QUANTUM THEORY

There are however exceptions to the Dulong and Petit law. For example at ordinary temperatures, diamond has a thermal capacity per gram-atom of only $6.3 \, \text{J K}^{-1}$. The reason for this is that the equipartition of energy is only valid if classical mechanics can be applied. For certain substances like diamond, quantum effects must be taken into account.

To develop a quantum theory of specific heats, a solid containing N atoms is again treated as a collection of N three-dimensional oscillators with the same frequency. The laws of quantum mechanics are now applied to these oscillators.

Using quantum mechanics, we can show that the energy of a one-dimensional harmonic oscillator has the possible values

$$E_n = (n + \tfrac{1}{2})h\nu$$

[3] It is always the thermal capacity at constant pressure C_p that is measured directly. The difference $C_p - C_V$ can be calculated by applying the relations of section 4.3 to the solid; however this difference is relatively small for a solid.

v is the natural frequency of the oscillator, h is Planck's constant ($h = 6.63 \times 10^{-34}$ J s) and n is zero or a positive integer.

According to the Boltzmann formula, the probability of an oscillator being in a state of energy E_n is

$$P(E_n) = A\,\mathrm{e}^{-E_n/kT} = A\,\mathrm{e}^{-(n+\frac{1}{2})hv/kT}$$

The mean energy of the oscillator is

$$\bar{E} = \sum_{n=0}^{\infty} E_n P(E_n) = A \sum_n (n + \tfrac{1}{2})hv\,\mathrm{e}^{-(n+\frac{1}{2})hv/kT}$$

The constant A is fixed by the normalization condition

$$\sum_n P(E_n) = A \sum_n \mathrm{e}^{-(n+\frac{1}{2})hv/kT} = 1$$

Putting $hv/kT = x$, we find

$$\bar{E} = \frac{hv \sum_n (n + \tfrac{1}{2})\,\mathrm{e}^{-(n+\frac{1}{2})x}}{\sum_n \mathrm{e}^{-(n+\frac{1}{2})x}} = hv \left[\frac{1}{2} + \frac{\sum_n n\,\mathrm{e}^{-nx}}{\sum_n \mathrm{e}^{-nx}} \right]$$

In order to calculate the quotient of the two series we write

$$\frac{\sum_n n\,\mathrm{e}^{-nx}}{\sum_n \mathrm{e}^{-nx}} = -\frac{d}{dx} \log \left[\sum_{n=0}^{\infty} (\mathrm{e}^{-x})^n \right] = -\frac{d}{dx} \log \frac{1}{1 - \mathrm{e}^{-x}}$$

$$= \frac{d}{dx} \log (1 - \mathrm{e}^{-x}) = \frac{\mathrm{e}^{-x}}{1 - \mathrm{e}^{-x}} = \frac{1}{\mathrm{e}^{x} - 1}$$

Finally, the internal energy of the solid is

$$U = 3N\bar{E} = 3Nhv \left[\frac{1}{2} + \frac{1}{\mathrm{e}^{hv/kT} - 1} \right]$$

In the high temperature limit when $hv/kT \ll 1$ we have,

$$\mathrm{e}^{hv/kT} - 1 \sim hv/kT$$

Neglecting the term $\frac{1}{2}$ compared with kT/hv we find

$$U \sim 3NkT$$

and

$$C_V = \frac{dU}{dT} \sim 3Nk$$

120

We have already obtained these last two results from classical mechanics. For the majority of solids the condition $hv/kT \ll 1$ is already satisfied at ordinary temperatures and so they obey the Dulong and Petit law.

However, at low temperatures U decreases more rapidly than $3NkT$ and C_V decreases. We can understand this by considering the extreme case $hv/kT \gg 1$. Then we can neglect 1 compared with $e^{hv/kT}$ and obtain

$$U \sim 3Nhv[\tfrac{1}{2} + e^{-hv/kT}]$$

and

$$C_V = \frac{dU}{dT} \sim 3Nk\left(\frac{hv}{kT}\right)^2 e^{-hv/kT}$$

It can be seen that the thermal capacity tends to zero at low temperatures and the Dulong and Petit law ceases to be valid. For a very hard solid such as diamond, the atoms experience large restoring forces and their natural frequency is high; at ordinary temperatures the condition $hv/kT \ll 1$ is already no longer satisfied and the thermal capacity is lower than 25 J K^{-1}.

We can sum up this section with the statement that, if the temperature is sufficiently high, the thermal capacity of a gram-atom of a solid is 25 J K^{-1} but this thermal capacity decreases and tends to zero at lower temperatures.

The theory just described, which is due to *Einstein*, does not correctly account for the quantitative behaviour of C_V at low temperatures however. Einstein's theory predicts an exponential decrease, whereas in fact C_V varies as T^3 at low temperatures. This shortcoming of Einstein's theory arises because it does not take into account all the interactions between the atoms in the solid. A later theory due to *Debye* does take account of these interactions and correctly explains the behaviour of C_V. This more complicated theory will not be treated here.

5.7 Specific heats of perfect gases

MONATOMIC GASES

We have already seen that the internal energy and the thermal capacity at constant volume of a monatomic perfect gas consisting of N molecules are respectively

$$U = \tfrac{3}{2}NkT \quad \text{and} \quad C_V = \frac{dU}{dT} = \tfrac{3}{2}Nk$$

This agrees with the principle of equipartition of energy. Each molecule has three degrees of freedom since its energy depends on the three velocity components. The energy is a quadratic function of these variables and each molecule has a mean energy equal to $\tfrac{3}{2}kT$.

DIATOMIC GASES

The energy of a diatomic molecule depends not only on the three velocity components but also on parameters which characterize its internal state. This internal state can be described by saying that the molecule can rotate and vibrate.

The rotational energy is in principle a quadratic function of the three components of angular momentum. In fact the rotational energy about the axis of the molecule is negligible for quantum-mechanical reasons. Thus the rotational energy effectively depends only on two components of angular momentum and the molecule has two rotational degrees of freedom.

The molecule can vibrate internally, the distance between the two atoms oscillating about a mean value. The corresponding energy is the same as that of a one-dimensional harmonic oscillator; it is a quadratic function of the separation and of the frequency of vibration. Thus the molecule has two vibrational degrees of freedom.

In all, the molecule has three translational, two rotational, and two vibrational degrees of freedom, making a total of seven. When classical mechanics is valid, the principle of equipartition of energy says that the internal energy and the thermal capacity at constant volume of a diatomic perfect gas composed of N molecules are respectively

$$U = \tfrac{7}{2}NkT \quad \text{and} \quad C_V = \frac{\mathrm{d}U}{\mathrm{d}T} = \tfrac{7}{2}Nk$$

In practice these values calculated using classical mechanics are only reached at very high temperatures. At ordinary temperatures for very many gases, the natural frequency of vibration v is such that $hv/kT \gg 1$. As we have already seen in connection with solids, the vibrational energy is negligible under these conditions and the two vibrational degrees of freedom do not contribute to the internal energy or the specific heat. This is why at ordinary temperatures the thermal capacity at constant volume is given approximately by

$$C_V = \tfrac{5}{2}Nk$$

At low temperatures, the rotational energy is not correctly described by classical mechanics. Its contribution to the specific heat diminishes and tends to zero, leaving only the three translational degrees of freedom. For example below 40 K the thermal capacity of molecular hydrogen at constant volume is $\tfrac{3}{2}Nk$, the same as for a monatomic gas. The details of the quantum-mechanical calculation of the rotational energy will not be given here.

Problems

5.1 N particles are distributed between two energy levels ε_1 and ε_2 where $\varepsilon_1 < \varepsilon_2$.

Sketch qualitatively the graph of the mean energy $\bar{E}$ of the system as a function of the temperature T. What is the value of $\bar{E}$ (a) when T is very large, (b) when T is very small? What is the transition temperature?

5.2 Perrin's experiment. Spherical particles of known radius a (of the order of 1 micron) and density ρ are suspended in liquid whose density ρ_0 is a little less than ρ. What is the distribution of particles in the liquid as a function of their height? Show that Avogadro's number can be deduced from this experiment.

5.3 Two types of macromolecules 1 and 2 with masses M_1 and M_2 are in suspension in a centrifuge which is rotating with an angular velocity ω. Determine the distribution of the molecules 1 and 2 as a function of the distance ρ from the axis of rotation at a temperature T.

How can the apparatus be used to separate molecules 1 and 2? Taking $M_1 = 90\,000$, $M_2 = 100\,000$ and assuming the centrifuge has a radius of 0.1 m and gives a maximum acceleration of $100\,000$ g, what is the maximum concentration of molecule 1 compared with molecule 2 that can be obtained in one operation?

5.4 A strip of elastomer can be considered as a single macromolecule consisting of N sections of length a oriented in any way whatever in space. The number of possible orientations within a solid angle $d\Omega$ is proportional to $d\Omega$. What is the length of strip as a function of the applied tension τ? Neglect interactions between different sections.

5.5 In the classical theory of paramagnetism (due to Langevin), it is assumed that the atomic magnetic moments can be oriented in *all* directions in space. The number of states oriented in a direction within a solid angle $d\Omega$ is proportional to $d\Omega$. Calculate the magnetization as a function of B and T. What is the susceptibility for a weak field?

5.6 The magnetic moment of an atom is of the order of 10^{-23} JT^{-1}. What magnetic field must be applied if five times as many atoms are to be aligned parallel to the field as antiparallel at the temperature of liquid nitrogen (77 K)?

5.7 For a gas consisting of molecules obeying the Maxwell distribution, calculate the mean values of the following quantities

$$v, v_x^2, v_x, v^2 v_x, v_x v_y$$

5.8 How many molecules in a mole of nitrogen have a speed greater than $1\,000\,\text{ms}^{-1}$ at a temperature of (a) 50 K, (b) 500 K, (c) 5 000 K?

5.9 Atomic beam. An oven at a temperature of $T = 1\,300\,\text{K}$ contains a gas of silver atoms at a pressure of $p_0 = 1\,\text{mm Hg}$. There is a slit measuring 10 mm by 0·05 mm in the wall of the oven, through which the atoms can pass into an evacuated enclosure that is continuously pumped. How many atoms of silver leave the oven per second?

In the wall of the enclosure at a distance $D = 1\,\text{m}$ from the slit, there is a second identical slit parallel to the first, through which atoms may pass into a second enclosure which is initially evacuated but not pumped and is kept at a temperature of 300 K. How many atoms per second pass through the second slit into this second enclosure? When equilibrium is established, what is the pressure in the second enclosure?

5.10 What is the rate of evaporation of a spherical drop of water of radius r at 100°C in an evacuated vessel which is continuously pumped?

5.11 An evacuated vessel of volume $V = 10^{-3}\,\text{m}^3$ is surrounded by air at a pressure of 1 atm and a temperature of 300 K. If a small hole of area $S = 10^{-10}\,\text{m}^2$ is made in the vessel, how long does it take the pressure inside to rise to $p = 1\,\text{mm Hg}$?

5.12 Cryogenic pumping. A vessel is initially full of water vapour at a pressure of $p_0 = 0·1\,\text{mm Hg}$ and a temperature $T = 300\,\text{K}$. A surface area $S = 10^{-4}\,\text{m}^2$ of the wall is cooled to liquid nitrogen temperature (77 K). At this temperature the saturated vapour pressure of water is negligible and every molecule striking the wall sticks to the surface. How long does it take the pressure to fall to $10^{-6}\,\text{mm Hg}$ in the vessel? Take the volume of the vessel as $10^{-3}\,\text{m}^3$.

5.13 The molecules $U_{235}F_6$ and $U_{238}F_6$ can be separated by diffusion through a porous partition. In natural uranium, the percentage of U_{238} is $c_1 = 99·3$ per cent and that of U_{235} is $c_2 = 0·7$ per cent. What is the ratio of the concentrations c'_2/c'_1 after diffusion into a vacuum. How many times must the process be repeated to raise the concentration of U_{235} to 50 per cent?

5.14 An atom emits radiation of frequency v_0 in its own rest frame. If this atom is moving at a speed v_z with respect to an observer, the observer measures the frequency as $v = v_0 (1 + v_z/c)$. What is the mean frequency observed from hydrogen at 10 000 K? If the spread in the frequency of the observed radiation is given by $\Delta v = \sqrt{[(v - \bar{v})^2]}$, derive an expression for the spread showing that the temperature of the gas may be measured from observations of the spread.

5.15 A vessel consists of two parts maintained at two different temperatures T_1 and T_2 connected by a pipe. The vessel contains a gas at pressure so low that there are collisions only between the wall and the molecules and not between the molecules themselves. Show that in each part the pressures p_1 and p_2 are different and obey the relation

$$\frac{p_1}{\sqrt{T_1}} = \frac{p_2}{\sqrt{T_2}}$$

5.16 The resistivity of a metal is due to the collisions between the electrons and thermally excited atoms of the metal. Assuming that the probability of collision of the electrons is proportional to the square of the amplitude of vibration of these atoms, determine the variation of resistance as a function of temperature.

5.17 If a mass dm is hung on a spring maintained at a temperature T, the spring extends by $dx = k\,dm$. What is the limit of the measurable mass dm due to thermal fluctuations?

5.18 A molecule of a perfect gas consists of four atoms situated at the corners of a tetrahedron. What are the number of translational, vibrational, and rotational degrees of freedom of this molecule? Hence deduce the specific heats of this gas at constant volume and at constant pressure.

5.19 The energy W of an ultra-relativistic particle such as a photon is related to its momentum $\mathbf{p}$ by the expression $W = |\mathbf{p}|c$ where c is the speed of light. Assuming, as in non-relativistic physics, that the number of states whose momentum vector points into a solid angle $d\Omega$ is proportional to $d\Omega$, calculate the internal energy of a gas of photons as a function of its temperature and hence its specific heat at constant volume.

Appendix I. Sound

We will study the propagation of sound, using the relation between the pressure and the volume in an adiabatic change established in section 2.8. We will see how the knowledge of this relation allows us to calculate the speed of sound.

The phenomenon of sound is caused by vibrations in a material medium. These vibrations occur on a scale large compared with intermolecular distances so that a macroscopic description is adequate; we can characterize the state of the medium by the displacement of each volume element, as if the medium were continuous. We will limit ourselves to fluids, where the vibrations are longitudinal, i.e., the displacements take place parallel to the direction of propagation.

We take the direction of propagation as the x-axis. A cross-section of the fluid which has the abscissa x at rest oscillates in time and has the abscissa $x + \psi$ at time t. The displacement ψ is a function of x and the time t. We intend to establish the partial differential equation which $\psi(x, t)$ obeys.

The mechanism of propagation is as follows. The displacements ψ generate variations in the density $\Delta\rho$; the density variations are accompanied by pressure variations Δp, which in turn produce other displacements ψ, so that a wave is propagated. ψ, Δp, and $\Delta\rho$ are small quantities. Thus for a sound of average intensity to the ear, Δp is of the order of 10^{-7} atm in air. Thus we can neglect higher order terms.

First let us relate ψ and $\Delta\rho$. Referring to Fig. A.1, we consider the section of fluid, which at rest is bounded by planes with abscissae x and $x + dx$. These planes move to $x + \psi(x, t)$ and $x + dx + \psi(x + dx, t)$ respectively. The thickness of the section changes from dx to

$$dx + \psi(x + dx, t) - \psi(x, t) = dx + \frac{\partial\psi}{\partial x}dx$$

127

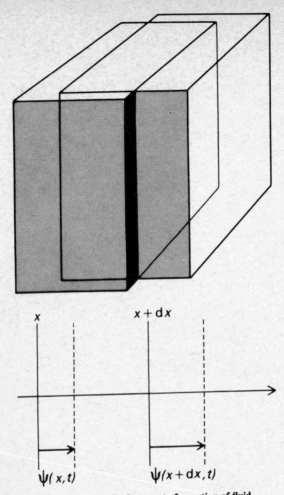

Figure A. 1. The displacement of a section of fluid.

In other words, the volume is multiplied by

$$1 + \frac{\partial \psi}{\partial x}$$

and the density at rest, ρ_0, is divided by this factor:

$$\rho_0 + \Delta\rho = \frac{\rho_0}{\left(1 + \dfrac{\partial \psi}{\partial x}\right)} \approx \rho_0 \left(1 - \frac{\partial \psi}{\partial x}\right)$$

Thus

$$\frac{\partial \psi}{\partial x} = -\frac{1}{\rho_0}\Delta\rho \qquad (1)$$

Now let us relate $\Delta\rho$ and Δp. The relation between the density and the pressure cannot be determined without an extra condition, such as a constant temperature for example. In fact this condition would not be correct. The oscillations are too rapid for heat to be transmitted from one section to another, as would be necessary to equalize the temperatures. On the contrary it is a good approximation to consider that there is no exchange of heat and that the compression of each section happens in an adiabatic manner. Furthermore, although too rapid to allow the exchange of heat, the oscillations are slow enough so that the compression can be considered reversible. Under these conditions, the correct relation to use to connect Δp and $\Delta\rho$ is the adiabatic reversible relation studied in section 2.8. Thus we have

$$\Delta p \approx \left(\frac{dp}{d\rho}\right)_{ad}\Delta\rho \qquad (2)$$

The derivative $dp/d\rho$ must be calculated from the adiabatic relation between p and ρ with the equilibrium conditions (p_0, ρ_0).

Finally let us relate Δp and ψ. A cylinder of fluid (Fig. A.2) of cross-sectional area S, between the planes with abscissae x and $x + dx$, has exerted on its ends the pressures

$$p_0 + \Delta p(x, t) \quad \text{and} \quad p_0 + \Delta p(x + dx, t)$$

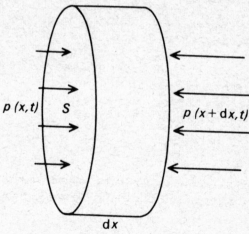

$$p\,(x,t) \quad S \qquad p\,(x + dx, t)$$

$$dx$$

Figure A.2. The forces on a cylinder of fluid.

respectively. Thus the net force on it is

$$S\Delta p(x, t) - S\Delta p(x + dx, t) = -S\frac{\partial(\Delta p)}{\partial x}dx$$

The volume of the cylinder is $S\,dx$ and its mass is $\rho_0 S\,dx$, neglecting the variation $\Delta\rho$ which will introduce only second order terms into the equation to follow. According to the fundamental equation of dynamics we have

$$-S\frac{\partial(\Delta p)}{\partial x}dx = \rho_0 S\,dx\frac{\partial^2\psi}{\partial t^2}$$

or

$$\frac{\partial^2\psi}{\partial t^2} = -\frac{1}{\rho_0}\frac{\partial(\Delta p)}{\partial x} \tag{3}$$

We now combine (1), (2), and (3):

$$\frac{\partial^2\psi}{\partial t^2} = -\frac{1}{\rho_0}\frac{\partial(\Delta p)}{\partial x} = -\frac{1}{\rho_0}\left(\frac{dp}{d\rho}\right)_{ad}\frac{\partial(\Delta p)}{\partial x} = \left(\frac{dp}{d\rho}\right)_{ad}\frac{\partial^2\psi}{\partial x^2}$$

Thus ψ obeys a wave equation

$$\frac{\partial^2\psi}{\partial t^2} = c^2\frac{\partial^2\psi}{\partial x^2}$$

where

$$c = \sqrt{\left(\frac{dp}{d\rho}\right)_{ad}}$$

The equation has solutions of the form $\psi = f(x \pm ct)$ where c is the velocity of propagation of the wave, i.e., the speed of sound in this case.

Thus we have expressed the speed of sound in terms of the derivative $dp/d\rho$, calculated for the medium concerned under adiabatic conditions.

We will complete the calculation for the particular case of a perfect gas, which will be an approximation to the case of a real gas. The relation between p and ρ is then $p = A\rho^\gamma$ so that

$$\left(\frac{dp}{d\rho}\right)_{ad} = \gamma A\rho^{\gamma-1} = \gamma\frac{p}{\rho}$$

p and ρ must be taken under equilibrium conditions for which the equation of state of a perfect gas is valid. For a mole of mass M, we have

$$p = \frac{RT}{V} = \rho\frac{RT}{M}$$

and thus

$$\left(\frac{dp}{d\rho}\right)_{ad} = \gamma\frac{p}{\rho} = \gamma\frac{RT}{M}$$

(under isothermal conditions, $dp/d\rho$ would simply be RT/M).

Thus the speed of sound is

$$c = \sqrt{\gamma\frac{RT}{M}}$$

It should be observed that $\frac{3}{2}RT$ is the kinetic energy of translation of the molecules in a mole:

$$\tfrac{3}{2}RT = \tfrac{1}{2}M\overline{v^2}$$

It can be seen that the speed of sound ($332\,\mathrm{ms}^{-1}$ in air at $0°C$) is of the order of the molecular speed $\sqrt{\overline{v^2}}$.

This is only to be expected since sound is propagated by the molecules colliding with one another.

Appendix II. Mean values

We recall here how mean values are obtained by the theory of probabilities.

Let us consider a variable E, which can take discrete values $E_1, E_2, E_3, \ldots$ with respective probabilities $P(E_1), P(E_2), P(E_3), \ldots$. This means that if E is measured a large number of times N, the value E_1 will occur $NP(E_1)$ times, the value E_2 will occur $NP(E_2)$ times and so on.

The total number of measurements is

$$N = NP(E_1) + NP(E_2) + \cdots$$

and $P(E_i)$ must therefore obey the normalization condition

$$\sum_i P(E_i) = 1$$

The mean value $\bar{E}$ is by definition the sum of all the values occurring divided by N, i.e.,

$$\bar{E} = \frac{NP(E_1) + NP(E_2) + \cdots}{N} = \sum_N E_i P(E_i)$$

These ideas can easily be extended to a variable q, which can have any value in the interval (q_1, q_2). The probability that the variable has a value between q and $q + dq$ is denoted by $f(q)\,dq$. The normalization condition is now

$$\int_{q_1}^{q_2} f(q)\,dq = 1$$

The mean value $\bar{q}$ is again obtained by multiplying every possible value q by the probability $f(q)\,dq$ and summing over all the possible values; here the sum is an integral and we have

$$\bar{q} = \int_{q_1}^{q_2} q f(q)\,dq$$

Appendix III. The calculation of some integrals

We will often need the integral

$$I = \int_{-\infty}^{\infty} e^{-\alpha x^2} \, dx$$

and integrals of a similar form.

I can be obtained by writing

$$I^2 = \int_{-\infty}^{\infty} e^{-\alpha x^2} \, dx \int_{-\infty}^{\infty} e^{-\alpha y^2} \, dy = \iint_{-\infty}^{\infty} e^{-\alpha(x^2 + y^2)} \, dx \, dy$$

Transforming to polar coordinates we find

$$I^2 = \int_{0}^{\infty} e^{-\alpha r^2} 2\pi r \, dr = \frac{\pi}{\alpha} \int_{0}^{\infty} e^{-\alpha r^2} \, d(\alpha r^2) = \frac{\pi}{\alpha}$$

where we have used the integral

$$\int_{0}^{\infty} e^{-t} \, dt = 1$$

Finally

$$I = \int_{-\infty}^{\infty} e^{-\alpha x^2} \, dx = \sqrt{\frac{\pi}{\alpha}}$$

By differentiation with respect to α, we obtain

$$\int_{-\infty}^{\infty} x^2 e^{-\alpha x^2} \, dx = -\frac{dI}{d\alpha} = \frac{1}{2} \frac{\sqrt{\pi}}{\alpha^{3/2}}$$

$$\int_{-\infty}^{\infty} x^4 e^{-\alpha x^2} \, dx = \frac{d^2 I}{d\alpha^2} = \frac{1}{2} \cdot \frac{3}{2} \frac{\sqrt{\pi}}{\alpha^{3/2}} \qquad \text{and so on.}$$

133

Since the integrand is even, these integrals taken between the limits 0 and $+\infty$ have half the values given above.

If the integrand contains an odd power of x, the result is even simpler. We have for example

$$\int_0^\infty e^{-\alpha x^2} x \, dx = \frac{1}{2\alpha} \int_0^\infty e^{-\alpha x^2} \, d(\alpha x^2) = \frac{1}{2\alpha}$$

and by differentiation with respect to α

$$\int_0^\infty e^{-\alpha x^2} x^3 \, dx = \frac{1}{2\alpha^2} \qquad \text{and so on.}$$

Appendix IV. Stirling's formula

We want to find the asymptotic form of $\log(n!)$ when n is large. We start from the integral representation

$$n! = \int_0^\infty t^n e^{-t} dt$$

This formula can be proved by differentiation with respect to α

$$\int_0^\infty t^n e^{-\alpha t} dt = (-1)^n \frac{d^n}{d\alpha^n} \int_0^\infty e^{-\alpha t} dt = (-1)^n \frac{d^n}{d\alpha^n}\left(\frac{1}{\alpha}\right) = \frac{n!}{\alpha^{n+1}}$$

and then putting $\alpha = 1$.

Thus we have

$$n! = \int_0^\infty t^n e^{-t} dt = \int_0^\infty e^{n \log t - t} dt$$

Successive differentiation of the exponent $y(t) = n \log t - t$ yields

$$y'(t) = \frac{n}{t} - 1$$

$$y''(t) = \frac{n}{t^2} \qquad \text{and so on}$$

$y(t)$ has a maximum when $t = n$, where $y'(t)$ is zero. If we develop $y(t)$ as a Taylor series about this point we obtain

$$y(t) = y(n) + y'(n)(t - n) + \tfrac{1}{2}y''(n)(t - n)^2 + \dots$$

$$= n \log n - n - \frac{1}{2n}(t - n)^2 + \dots$$

If we put $t - n = u$, the integral to be calculated becomes

$$n! = \int_{-n}^{\infty} e^{n \log n - n - (1/2n)u^2 + \cdots} \, du$$

If n is very large, the lower limit of the integral can be replaced by $-\infty$. Then we have

$$n! \sim e^{n \log n - n} \int_{-\infty}^{\infty} e^{-u^2/2n} \, du = e^{n \log n - n} \sqrt{2\pi n}$$

(the integral is calculated in Appendix III). Finally

$$\log(n!) \sim n \log n - n + \tfrac{1}{2} \log(2\pi n)$$

It is often sufficient to write

$$\log(n!) \sim n \log n - n$$

The search for a more rigorous treatment is left to the mathematically minded reader.

Appendix V. Partial derivatives

We do not intend to take up the study of partial derivatives, but merely wish to emphasize the fact that they are only defined if the choice of independent variables has been made. We will content ourselves with an example.

Let us consider the internal energy U of a gram-atom of a monatomic gas using Van der Waals' approximation. This energy as a function of T and V is

$$U = \frac{3}{2}RT - \frac{A}{V}$$

The equation of state

$$p = \frac{RT}{V - B} - \frac{A}{V^2}$$

relates the three variables pressure p, volume V, and temperature T, so that the energy may be written as a function of p and V:

$$U = \frac{3}{2}p(V - b) + \frac{1}{2}\frac{A}{V} - \frac{3}{2}\frac{AB}{V^2}$$

If the first expression for U is differentiated with respect to V, we find

$$\frac{\partial U}{\partial V} = \frac{A}{V^2}$$

If the second form of U is differentiated with respect to V, a different expression is found:

$$\frac{\partial U}{\partial V} = \frac{3}{2}p - \frac{1}{2}\frac{A}{V^2} + \frac{3AB}{V^3}$$

It can be seen that $\partial U/\partial V$ means nothing unless the other independent variable is specified. The correct formulae are

$$\left(\frac{\partial U}{\partial V}\right)_T = \frac{A}{V^2}$$

and

$$\left(\frac{\partial U}{\partial V}\right)_P = \frac{3}{2}p - \frac{1}{2}\frac{A}{V^2} + \frac{3AB}{V^3}$$

Bibliography

The very limited list of references given here is of necessity extremely arbitrary. The level of most of the treatments indicated is higher than that of a first-year undergraduate course although a number of excellent books have been omitted because they are too difficult.

1. Fundamental treatments

F. Reif, *Statistical Physics, Berkeley Physics Course*, Vol. 5, McGraw-Hill, New York, 1967.

This book deserves a special place in this bibliography since like the present work it endeavours to present a unified account of statistical physics and thermodynamics. While *Statistical Physics* gives only a brief description of physical phenomena, it treats the theoretical aspects on a more sophisticated level than the present text. Considerable care has been taken in the presentation of the material. *Statistical Physics* will be particularly useful to those who wish to develop their understanding of the fundamentals of the subject.

R. P. Feynman, *Lectures on Physics*, Vol. 1, Addison-Wesley, Reading, 1963.

This well-known text contains several chapters devoted to thermodynamics and statistical physics. It requires little technical knowledge of the reader and is written principally for newcomers to the subject with the intention of generating enthusiasm for physics among those potentially capable of it. In fact even experienced physicists do not tire of reading it again and again.

F. Mandl, *Statistical Physics*, Wiley, London, 1971.

This excellent introductory text follows the same general pattern as the present book, but at a somewhat higher level.

2. Traditional presentations

The reader is strongly recommended to study a macroscopic treatment of classical thermodynamics in parallel with or immediately after this book.

An elementary account is given in

M. W. Zemansky, *Heat and Thermodynamics*, 4th edn., McGraw-Hill, New York, 1957.

Rather condensed accounts at a higher level can be found in

E. Fermi, *Thermodynamics*, Dover Publications, New York, 1957.
A. B. Pippard, *Elements of Classical Thermodynamics*, Cambridge University Press, London, 1964.

An account of classical thermodynamics followed by a treatment of statistical physics is contained in

F. W. Sears, *Thermodynamics*, Addison-Wesley, Reading, 1953.

3. Statistical physics

A relatively elementary and extremely clear account is to be found in

J. D. Fast, *Entropy*, Phillips, 1962.

4. Texts written specifically for first-year undergraduate courses

Several chapters devoted to thermodynamics can be found in,

F. W. Sears and M. W. Zemansky, *University Physics: Part 1*, 3rd edn., Addison-Wesley, Reading, 1964.
D. Halliday and R. Resnick, *Fundamentals of Physics*, Wiley, New York, 1970.

Answers to the problems

Chapter 1

1.1 $5.936 \times 10^{-3} \, m^3$; $0.986 \times 10^{-5} \, N^{-1} \, m^2$; $3.658 \times 10^{-3} \, K^{-1}$; $3.661 \times 10^{-3} \, K^{-1}$

1.2 $\alpha = p\chi\beta$

1.3 $p(V - b) = rT$

1.4 Helium

1.5 $V_c = 2b$; $p_c = a/4e^2b^2$; $RT_c = a/4b$.

1.6 $A = rT$; $B = b - a/rT$; $C = 2ab/r^2T^2 - a^2/r^3T^3$

1.7 831; 1 662 kg; 527 atm; $534 \times 10^5 \, N \, m^{-2}$

1.8 48 kg; 288 m^3

1.10 486 ms^{-1}

1.11 25.9°C; 37.5°C

1.12 50.25°C

1.13 $a = 0.39882 \times 10^{-2}$; $b = -6.88 \times 10^{-7}$; 162.5°C

1.14 2.41 atm; 3.41 atm

1.15 0.55 atm; 86°C

1.16 $n_1 = 650$ mm; $n_1' = 750$ mm; $n_2 = 566$ mm; $n_2' = 666$ mm

1.17 all $+100$ mm

1.18 Unchanged

1.19 7.5 mm

Chapter 2

2.1 2.2°C

2.2 $25.0 \, J \, mole^{-1} \, K^{-1}$; $24.3 \, J \, mole^{-1} \, K^{-1}$

2.3 $0.7 \, J \, K^{-1}$; $5.6 \, J \, K^{-1}$; $52.5 \, J$; $2.63 \, J \, K^{-1}$

2.4 $c_p = c_V + r\left[\dfrac{1}{1 - 2a(V - b)^2/rTV^3}\right]$

2.5 $U = c_V T - \dfrac{a}{V}$

2.6 $\left(p + \dfrac{a}{V^2}\right)(V - b)^{r + c_V/c_V} = \text{Constant}$

2.7 $776 \, J \, K^{-1}$; $3.35 \times 10^5 \, J \, kg^{-1}$; $1.99 \, J \, kg^{-1}$

2.8 2 393 K; $48.9 \, J \, mole^{-1} \, K^{-1}$; $206 \, kJ \, mole^{-1}$

2.9 1 245 ohms

2.10 7.7×10^3 kg

2.11 377°C; 33·8 atm

2.12 90 m s^{-1}; 151 K; 1·05 atm

2.13 $\dfrac{dT}{dh} = -\dfrac{\gamma - 1}{\gamma}\dfrac{gM}{R} = 0.97 \times 10^{-2}$ K m^{-1}

where M = mean molecular weight of air.

2.14 $\Delta p_1 = \Delta p_0 \dfrac{\gamma - 1}{\gamma}$

2.15 $T_f = \gamma T_0$; $W = Q = p_0 V/\gamma$

2.16 45 s; 32 s $(\gamma = 1.4)$

2.17 (a) $p_A = p_B = p_C = 1.15$ atm; $T_B = T_C = 318$ K.

(b) Same answer as (a).

2.18 (a) $p_A = p_B = p_0^{2-\gamma}$; $T_A = T_B = T_0 2^{1-\gamma}$

(b) $p_A = p_B = p_0$; $T_A = T_B = T_0\left(2 - \dfrac{1}{\gamma}\right)$

(c) $p_A = p_B = p_0$; $T_A = T_0$; $T_B = \gamma T_0$

Chapter 3

3.1 $T = \dfrac{M_1 T_1 + M_2 T_2}{M_1 + M_2}$; $\Delta S = cM_1 \log\dfrac{T}{T_1} + cM_2 \log\dfrac{T}{T_2}$

3.2 $T = \dfrac{(p_1 + p_2)T_1 T_2}{p_1 T_2 + p_2 T_1}$; $p = \dfrac{(p_1 + p_2)}{2}$

$\Delta S = \left(\dfrac{p_1 V}{T_1} + \dfrac{p_2 V}{T_2}\right)\log 2 + \dfrac{c_V}{R}\left(\dfrac{p_1 V}{T_1}\log\dfrac{T}{T_1} + \dfrac{p_2 V}{T_2}\log\dfrac{T}{T_2}\right)$

3.3 $T = \dfrac{T_1 + T_2}{2}$; $p = \dfrac{p_1 V_1 + p_2 V_2}{V_1 + V_2}$

$\Delta S = RN \log\dfrac{(T_1/p_1 + T_2/p_2)^2}{T_1 T_2/p_1 p_2} + c_V N \log(T^2/T_1 T_2)$

3.4 3.8×10^5 J

3.5 5·7 kW

3.6 3·3 J K^{-1}; 3·3 J K^{-1}

3.7 (a) $\log\Omega = N \log 2N - \tfrac{1}{2}\left(N - \dfrac{U}{\mu B}\right)\log\left(N - \dfrac{U}{\mu B}\right)$

$\qquad\qquad - \tfrac{1}{2}\left(N + \dfrac{U}{\mu B}\right)\log\left(N + \dfrac{U}{\mu B}\right)$

(b) $U = -N\mu B \tanh(\mu B/kT)$

(c) $\dfrac{\mu B}{kT} = \tanh^{-1}(\mu B/kT) = $ Constant

3.8 $W_m = Mc(\sqrt{T_2} - \sqrt{T_1})^2$; $T_f = \sqrt{T_1 T_2}$

3.10 $W_m = Mc[T_1 - T_0 + T_0 \log(T_0/T_1)]$

3.11 $W_m = RT_0\left[\log(p_1/p_0) + p_{\text{ext}}\left(\dfrac{1}{p_1} - \dfrac{1}{p_0}\right)\right]$

3.12 $T_1 = T_0\left(\dfrac{p_1}{p_0}\right)^{\gamma - 1/\gamma}$; $\quad \Delta S_1 = R \log \dfrac{p_1}{p_0}$

$\quad\quad T_n = T_0\left(\dfrac{p_1}{p_0}\right)^{n(\gamma - 1)/\gamma}$; $\quad \Delta S_n = nR \log \dfrac{p_1}{p_0}$

3.14 $W = (V_2 - V_1)(p_2 - p_1)$

Chapter 4

4.1 -134 atm K^{-1}; 269 atm

4.2 $-0.2°C$

4.5 $5.3°C$

4.6 3.5

4.9 $\Delta T_1 = -\dfrac{T\lambda L\tau}{C}$; $\quad \Delta T_2 = \dfrac{\tau\lambda LT}{C}\left(1 + \dfrac{\tau\alpha}{2\lambda T}\right)$

where C = thermal capacity of the wire and $\lambda = \dfrac{1}{L}\left(\dfrac{\partial L}{\partial T}\right)_\tau$

4.11 (a) $T_2 = T_1\left(\dfrac{p_2}{p_1}\right)^{(\gamma - 1)/\gamma}$

$\quad\quad$ (b) As for (a)

$\quad\quad$ (c) $\dfrac{T_2}{T_1} = \left(\dfrac{p_2 + a/V_2^2}{p_1 + a/V_1^2}\right)^{(\gamma - 1)/\gamma}$

4.13 $\Delta U = -233 \times 10^3 \text{ J}$; $\Delta T = +23.7°C$

Chapter 5

5.1 $N(\varepsilon_1 + \varepsilon_2)/2$; $N\varepsilon_1$; $(\varepsilon_1 - \varepsilon_2)/k$

5.2 $\dfrac{dN}{dh} \propto \exp\left(-\dfrac{4\pi a^3(\rho - \rho_0)}{3}\dfrac{gh}{kT}\right)$

5.3 $\dfrac{dN_{1,2}}{dh} \propto \exp\left(-\dfrac{M_{1,2}\omega^2\rho^2}{2kT}\right)$

5.4 $L = Na\left(\coth\dfrac{\tau a}{kT} - \dfrac{kT}{\tau a}\right)$

5.5 $\mathcal{I} = N\mu\left(\coth\dfrac{\mu B}{kT} - \dfrac{kT}{\mu B}\right)$

5.6 86 teslas

5.7 $\left(\dfrac{8kT}{\pi m}\right)^{\frac{1}{2}}$; $\dfrac{kT}{m}$; $0; 0; 0$

5.8 9.7×10^9; 5.0×10^{22}; 5.3×10^{23}

5.9 $4.7 \times 10^{17} \text{ s}^{-1}$; $7.5 \times 10^{10} \text{ s}^{-1}$; $0.77 \times 10^{-7} \text{ mmHg}$

5.10 0.1 ms^{-1}

5.11 113 s

5.12 0.77 s

5.13 $\left(\dfrac{c_2}{c_1}\right)\left(\dfrac{m_1}{m_2}\right)^{\frac{1}{2}}$; about 500.

5.14 $\bar{v} = v_0$; $\Delta v = v_0 \sqrt{\dfrac{kT}{mc^2}}$

5.16 $\rho \propto T$

5.17 $\mathrm{d}m \sim \left(\dfrac{kT}{Kg}\right)^{\frac{1}{4}}$

5.18 18

Index

Set on Monophoto Filmsetter and printed by J. W. Arrowsmith Ltd., Bristol, England.